Generis

PUBLISHING

AF687594

Environmental Communication and Sustainable Rural Development in Algeria

Concepts, Policies, and Theoretical Foundations

Dr. Hocine Bousalah

Copyright © 2025 Dr. Hocine Bousalah
Copyright © 2025 Generis Publishing

All rights reserved. This book or any portion thereof may not be reproduced or used in any manner whatsoever without the written permission of the publisher except for the use of brief quotations in a book review.

Title: **Environmental Communication and Sustainable Rural Development in Algeria**

Concepts, Policies, and Theoretical Foundations

ISBN: 979-8-89966-250-8

Author: Dr. Hocine Bousalah

Cover image: www.unsplash.com

Publisher: Generis Publishing
Online orders: www.generis-publishing.com
Contact email: info@generis-publishing.com

Environmental Communication and Sustainable Rural Development in Algeria: Concepts, Policies, and Theoretical Foundations

Table of Contents

List of Tables

Chapter 1 – Introduction: Setting the Scene

1.1 Global Context of Environmental Communication

Environmental communication has emerged over the past four decades as a critical field at the intersection of communication studies, environmental sciences, and sustainable-development policy. It encompasses the study and practice of how societies construct, negotiate, and circulate meanings about nature, ecological risks, and sustainability challenges through media discourse, institutional messaging, and public dialogue (Cox & Pezzullo, 2021). In a world increasingly shaped by climate change, biodiversity loss, and socio-ecological inequalities, communication is no longer a peripheral dimension of environmental governance but rather a constitutive force in shaping public perception, collective behavior, and policy legitimacy.

At the international level, the 1992 Rio Earth Summit and subsequent global frameworks—such as the 2030 Agenda for Sustainable Development—have institutionalized the idea that effective communication and public participation are prerequisites for achieving sustainable development goals (United Nations Development Programme [UNDP], 2019; United Nations Environment Programme [UNEP], 2023). Scholars have increasingly emphasized the notion of *participatory sustainability*, where environmental issues are framed not merely as technical or scientific problems but as communicative processes rooted in culture, politics, and social power (Dryzek, 2020; Servaes, 2018). Within this paradigm, communication serves both as an informational instrument—transmitting knowledge about environmental degradation—and as a dialogical platform that enables citizens, experts, and policymakers to co-construct solutions.

In the Global South, and particularly within North African contexts, the challenge of environmental communication takes on a multidimensional character. It is embedded in postcolonial trajectories of development, institutional modernization, and environmental governance. Countries such as Algeria, Morocco, and Tunisia are confronted with the dual imperative of sustaining economic growth while safeguarding fragile ecosystems threatened by desertification, water scarcity, and rural poverty (FAO, 2022). Consequently, the integration of communication strategies into rural-development policies becomes a strategic necessity. Environmental awareness campaigns, community radio initiatives, and local associations have increasingly sought to translate global sustainability discourses into culturally relevant practices.

In Algeria, the intersection between environmental communication and rural development is particularly significant. The country's vast rural territories—stretching from the Tellian highlands to the Saharan fringes—constitute both the backbone of agricultural production and the locus of environmental vulnerability. The government's Integrated Rural Development Program (*Programme National de Développement Rural Intégré – PNDAR*) and subsequent sustainability strategies (notably the *Plan National de Développement Durable – PNDD*) have recognized communication as a cross-cutting mechanism for mobilizing local participation, disseminating ecological knowledge, and reinforcing behavioral change. However, despite notable progress, the communicative dimension of rural sustainability remains under-theorized in Algerian scholarship, which has often privileged economic or administrative approaches to development.

From a disciplinary standpoint, environmental communication in Algeria occupies a nascent yet promising position within media and communication studies. The integration of environmental themes in journalism, education, and institutional discourse is still fragmented, lacking a comprehensive theoretical synthesis that connects communication practice with sustainable-development policy. This book therefore seeks to fill that gap by providing an in-depth theoretical examination of how communication operates as a foundational pillar for achieving ecological modernization and sustainable rural transformation in Algeria and comparable Arab contexts.

1.2 The Problem and Rationale

Despite the centrality of environmental communication in global sustainability discourses, its theoretical and practical integration within developing societies—particularly in North Africa—remains limited and under-conceptualized. In many contexts, including Algeria, communication is still perceived as a secondary or instrumental function within environmental policy frameworks rather than as an essential dimension of governance, civic participation, and social transformation (Servaes, 2018; Pezzullo & Cox, 2021). This limited view often results in fragmented outreach programs and short-term awareness campaigns that fail to foster deep behavioral change or sustained public engagement.

At the heart of this problem lies a structural imbalance between *policy articulation* and *communicative practice*. Environmental initiatives in the Global South are frequently designed within centralized administrative frameworks, leaving little room for participatory processes that enable citizens—especially rural populations—to contribute their local knowledge and adaptive practices. As a result, communication remains confined to top-down dissemination models, emphasizing information transfer over dialogue and collective learning (Fairclough, 1995; Dryzek, 2020). The absence of participatory communication mechanisms undermines not only the social legitimacy of environmental policies but also their long-term effectiveness in promoting sustainable lifestyles and ecological stewardship.

In the Algerian case, this challenge is amplified by socio-economic and geographical factors. Algeria's vast and ecologically diverse territory—ranging from Mediterranean coastlines to arid Saharan zones—renders environmental management a complex, multi-scalar endeavor. Rural regions, in particular, experience acute environmental stress caused by deforestation, soil erosion, overgrazing, and the impacts of climate variability (Ministry of Environment and Renewable Energies, 2023). Yet, these same regions are also characterized by strong social cohesion, traditional ecological knowledge, and community-driven mechanisms that can serve as powerful entry points for environmental communication and education. The paradox, therefore, is that the areas most affected by environmental degradation often remain the least integrated into formal communication and awareness structures.

This theoretical gap raises critical questions: How can environmental communication evolve from a mere dissemination tool into a participatory instrument of social change? How can communication frameworks bridge the divide between expert-driven discourse and the lived realities of rural communities? And to what extent can communication be institutionalized as a pillar of sustainable rural development, rather than a peripheral activity attached to development projects?

Addressing these questions requires moving beyond descriptive or policy-oriented approaches toward a more systemic understanding of communication as a social process. This book argues that communication should be reconceptualized as a *constitutive dimension of sustainability*, not merely as an operational device for environmental advocacy. By engaging with theories of discourse, participation, and social learning, it seeks to reframe environmental communication within the broader epistemology of sustainable development and ecological modernity.

Moreover, the rationale for this work stems from the need to articulate a **contextually grounded framework** for environmental communication in Algeria—one that integrates global theoretical advances with the specific socio-cultural realities of the Arab-Maghreb region. While international literature has expanded significantly since the 1990s, with notable contributions from Western scholars, there remains a relative scarcity of Arabic-language research addressing the communicative dimensions of environmental governance. The present work, therefore, serves a dual purpose: to synthesize international theoretical perspectives and to situate them within Algerian and North African contexts through critical reinterpretation and comparative analysis.

Ultimately, this book proceeds from the conviction that sustainable development cannot be achieved without communicative sustainability. The capacity of societies to negotiate meanings about the environment—to debate, contest, and reimagine their relationship with nature—is as crucial as technological innovation or policy design. Environmental communication, understood as a field of power, discourse, and participation, offers a strategic pathway toward that transformation.

1.3 Objectives and Scope of the Book

The principal objective of this book is to establish a comprehensive theoretical framework that situates **environmental communication** as a cornerstone of **sustainable rural development**, with a specific emphasis on the Algerian and wider Arab–Maghreb contexts. Rather than limiting itself to empirical description or localized case studies, the book adopts a **conceptual and analytical perspective**, synthesizing global theories with regionally grounded realities. It seeks to contribute to the academic discourse on how communication, as a cultural, institutional, and participatory process, can shape environmental consciousness, influence policy, and foster community resilience.

More specifically, the book pursues the following interrelated objectives:

1. **To conceptualize environmental communication as a multidimensional discipline.**
   The book aims to move beyond its conventional treatment as an extension of environmental education or advocacy. It argues that environmental communication should be understood as an interdisciplinary field intersecting with sociology, ecology, political science, and media studies. Within this framework, communication is viewed as both a process of symbolic construction and a mechanism of social change (Cox & Pezzullo, 2021; Servaes, 2018).

2. **To integrate environmental communication within the paradigm of sustainable development.**
   Building upon the Sustainable Development Goals (SDGs) and contemporary ecological theories, the book explores how communication contributes to achieving environmental sustainability, social equity, and economic balance. It underscores that without effective communication strategies—rooted in participation, transparency, and knowledge sharing—sustainability efforts remain fragmented and elitist.
3. **To reinterpret rural development through the lens of communication.**
   The book redefines the notion of *rural development* not as a purely economic or agricultural process, but as a communicative and cultural transformation. It highlights how communication enables local populations to co-produce knowledge, negotiate priorities, and adapt traditional ecological wisdom to contemporary challenges such as climate change, water scarcity, and agricultural modernization (FAO, 2022).
4. **To bridge global theory and local experience.**
   A key ambition of this work is to reconcile global frameworks—rooted in Western academic traditions—with the specificities of North African societies. It calls for a decolonized understanding of environmental communication that accounts for local epistemologies, languages, and cultural dynamics. Through critical engagement with Arabic and Francophone sources, it seeks to reposition Algerian and Maghrebian perspectives within the global academic conversation.
5. **To propose a model for communicative sustainability in Algeria.**
   By the end of the book, a conceptual model will be developed that outlines the key components, actors, and strategies of environmental communication in Algeria. This model will integrate theoretical constructs such as participatory governance, ecological modernization, and communicative rationality (Habermas, 1984), offering a holistic vision for embedding communication within sustainable-development frameworks.

In terms of **scope**, this book is theoretical in nature but contextual in orientation. It does not attempt to replicate or update previous empirical fieldwork conducted in Algerian rural areas; rather, it builds upon such work to formulate a refined conceptual synthesis. The focus is on analyzing discourses, frameworks, and policies that shape environmental communication as a field of knowledge and practice in Algeria and comparable developing countries. Therefore, the geographic scope extends beyond Algeria to include comparative insights from the Maghreb, the Arab world, and the Global South.

Temporal boundaries of the analysis cover the period from the early 1990s—marked by the institutionalization of sustainable development after the Rio Earth Summit—to the present decade (2020s), characterized by digital transformation and the rise of climate communication. This temporal framing allows the book to trace both the evolution of environmental communication and its contemporary manifestations in digital, participatory, and policy-oriented domains.

Finally, the scope of this book encompasses both **academic and policy relevance**. Academically, it provides a synthesis of theories, frameworks, and perspectives that enrich the literature on environmental communication and sustainability studies. From a policy standpoint, it offers reflections that may inform decision-makers, media practitioners, educators, and NGOs seeking to integrate communication strategies into environmental governance and rural-development programs. The intersectional nature of this work—bridging theory, communication, and sustainability—positions it as a resource for scholars, students, and professionals interested in rethinking how societies communicate about their environment and their future.

1.4 Conceptual and Theoretical Orientation

The conceptual orientation of this book rests on the premise that **communication is not merely an auxiliary tool of environmental policy**, but a constitutive process that shapes how societies define, negotiate, and act upon sustainability challenges. It draws upon a diverse body of theoretical traditions—ranging from discourse theory and participatory communication to ecological modernization and environmental governance—to construct an integrated framework capable of explaining the complex relationship between communication, society, and environment in the Algerian and broader Global South contexts.

1.4.1 Communication as a Constitutive Process

At the most fundamental level, this book adopts a **constitutive view of communication**, as opposed to the traditional *transmission model*. The latter, dominant in administrative and development communication throughout the twentieth century, conceptualized communication as the linear transfer of information from sender to receiver. However, contemporary scholars such as Craig (1999), Habermas (1984), and Fairclough (1995) have demonstrated that communication is a **social**

process through which meaning, power, and identity are co-constructed. Within environmental contexts, this implies that ecological issues—such as deforestation, pollution, or climate change—are not purely biophysical phenomena but discursive constructs mediated by language, media, and institutional frameworks (Hajer, 1995; Cox & Pezzullo, 2021).

From this perspective, environmental communication is both a *discursive arena* and a *social practice*. It reflects how actors—governments, scientists, journalists, NGOs, and citizens—frame the environment, articulate responsibility, and define what counts as legitimate knowledge. The field therefore engages deeply with questions of ideology, representation, and voice, examining how communication either reinforces or challenges existing power structures. In this sense, discourse analysis becomes an essential methodological and epistemological lens through which the communicative dimensions of environmental governance can be critically explored (van Dijk, 2000; Dryzek, 2020).

1.4.2 Participatory Communication and Development

The second conceptual pillar of this book is **participatory communication**, rooted in the broader paradigm of **Communication for Development (C4D)**. Emerging in the 1970s as a critique of modernization theories, participatory communication emphasizes horizontal, dialogical, and inclusive approaches to social change (Freire, 1970; Servaes, 2018). In contrast to top-down dissemination models, it views communities not as passive recipients of information but as active agents capable of producing and validating their own knowledge.

Applied to environmental contexts, participatory communication enables local populations to engage in the diagnosis, planning, and monitoring of sustainability initiatives. It reinforces the notion of *environmental citizenship*, where communication acts as a channel for empowerment and collective action (Manyozo, 2012). For rural development, particularly in Algeria and North Africa, this approach resonates strongly with traditional communal practices such as *touiza* (collective work) and village assemblies, which historically embodied participatory decision-making. Recognizing these culturally embedded forms of communication can help build hybrid models that combine modern sustainability frameworks with indigenous modes of collaboration and environmental stewardship.

Participatory communication also provides a bridge between scientific knowledge and local experience. By valuing narrative, dialogue, and local storytelling, it allows for a more culturally sensitive and context-specific translation of environmental information. This, in turn, enhances the social legitimacy of environmental programs and policies.

1.4.3 Ecological Modernization and Environmental Governance

A third theoretical strand mobilized in this book is ecological modernization theory (EMT), which interprets environmental reform as a rational modernization process embedded within institutions, markets, and technological innovation (Mol & Spaargaren, 2000). According to this perspective, modern societies can achieve sustainability not by rejecting industrial development, but by transforming its underlying logics through innovation, efficiency, and participatory governance. However, this book adopts a critical ecological-modernization stance, emphasizing that technological solutions must be accompanied by communicative transformations—new narratives, inclusive dialogues, and transparent institutional practices—to ensure that modernization remains socially equitable and culturally legitimate.

Closely related to this is the framework of environmental governance, which underscores the role of communication in mediating relationships among stakeholders—state institutions, civil society, the private sector, and citizens (UNDP, 2023). Governance, in this sense, is not merely administrative management but a communicative process of coordination, negotiation, and accountability. It involves creating spaces—both physical and digital—where environmental knowledge can circulate freely, where stakeholders can deliberate openly, and where decisions are co-produced rather than imposed. Within Algeria's evolving governance landscape, this perspective invites rethinking the role of public media, education systems, and civil organizations in facilitating participatory dialogue on environmental issues.

1.4.4 Critical and Postcolonial Perspectives

While participatory and governance-based models provide normative foundations, they often emerge from Western epistemological frameworks. Therefore, this book also engages with critical and postcolonial perspectives to situate environmental communication within the socio-political realities of the Arab–Maghreb region.

Scholars such as Escobar (1995) and Said (1978) remind us that discourses of "development" and "modernization" are often entangled with histories of colonial domination and cultural asymmetry. In Algeria, for example, the legacy of colonial land policies, extractivist economies, and imported development paradigms continues to influence how environmental issues are defined and managed.

A postcolonial approach thus calls for epistemic pluralism—a recognition that environmental knowledge does not emanate solely from scientific institutions but also from local cultures, spiritual traditions, and community practices. It demands that communication scholarship incorporate indigenous perspectives, oral traditions, and collective memory as legitimate forms of ecological knowledge (Al-Dulaimi, 2012). This epistemological diversification enriches environmental communication, positioning it as a field capable of bridging global discourses with local lived realities.

1.4.5 Integrative Framework for Environmental Communication

Synthesizing these traditions, this book advances an integrative theoretical framework that conceptualizes environmental communication as:

1. A **discursive process**, through which meanings about nature and sustainability are constructed, contested, and disseminated;
2. A **participatory practice**, enabling inclusive dialogue and collective decision-making;
3. A **governance mechanism**, mediating relationships among institutions, policies, and citizens; and
4. A **cultural and postcolonial field**, rooted in diverse epistemologies and shaped by historical trajectories.

This theoretical orientation serves as the foundation for the chapters that follow. It allows the book to analyze the communicative dimensions of sustainability not merely as rhetoric but as a living, dynamic process deeply embedded in the political ecology of contemporary Algeria and the Arab world. By linking these frameworks, the book aspires to articulate a holistic vision of communicative sustainability, where dialogue, participation, and cultural diversity form the core of environmental transformation.

Chapter 1 – Introduction: Setting the Scene

This introductory chapter provides the conceptual and contextual grounding for the book. It situates environmental communication within global sustainability discourses and outlines the problem, rationale, objectives, theoretical orientation, and structure of the work. The chapter also justifies the need for a re-theorization of communication as a constitutive process in achieving sustainable rural development.

Chapter 2 – Theoretical Foundations of Environmental Communication

This chapter delves into the historical and epistemological evolution of environmental communication as a field. It discusses its origins in communication theory, environmental sociology, and discourse analysis. It explores major schools of thought—from positivist and functionalist models to critical and participatory approaches—and situates environmental communication within the broader intellectual landscape of media and environmental studies.

Chapter 3 – Communication, Sustainability, and Development Paradigms

Here, the focus shifts toward the relationship between communication and sustainable development. The chapter analyzes conceptual frameworks such as the Sustainable Development Goals (SDGs), ecological modernization, and communication for development (C4D). It also critiques the limitations of modernization-based approaches and advances an integrated model linking communication to environmental ethics, cultural sustainability, and participatory governance.

Chapter 4 – Rural Development and Environmental Policies in Algeria

This chapter contextualizes the discussion within Algeria's socio-political and ecological landscape. It traces the evolution of rural and environmental policies—from the post-independence agricultural reforms to contemporary sustainability strategies such as the *PNDAR* and *PNDD*. It highlights how communication practices intersect with state policy, media initiatives, and community-based programs in addressing environmental challenges.

Chapter 5 – The Role of Media and Communication Actors

This chapter examines the roles of various communication agents—journalists, educators, NGOs, and public institutions—in shaping environmental discourse. It evaluates how traditional and digital media contribute to public awareness, policy framing, and social mobilization. Comparative insights from the Maghreb and broader

Arab world will illustrate how communicative infrastructures can either advance or constrain environmental transformation.

Chapter 6 – Towards a Model of Communicative Sustainability in Algeria

The final chapter synthesizes insights from the previous sections to propose a conceptual model for *communicative sustainability*. It outlines guiding principles for integrating environmental communication into policy, education, and civic engagement. This model aspires to serve as a theoretical and practical framework for scholars, practitioners, and policymakers seeking to embed communication within the fabric of sustainable development.

Chapter 2 – Theoretical Foundations of Environmental Communication

2.1 Historical Emergence of Environmental Communication

2.2 Defining the Field: Concepts and Boundaries

The academic and practical field of environmental communication emerged gradually in the latter half of the twentieth century, driven by the growing recognition that ecological crises are not merely scientific or technological problems but also communicative, cultural, and political phenomena.

Understanding how societies talk about, represent, and respond to environmental challenges became essential for shaping both policy and public awareness.

2.1.1 Early Environmental Awareness and the Role of Media (1960s–1980s)

The intellectual roots of environmental communication can be traced back to the environmental movements of the 1960s and 1970s, particularly in the United States and Western Europe. The publication of Rachel Carson's *Silent Spring* (1962) marked a defining moment by exposing the ecological dangers of pesticides and awakening the public conscience to the interconnectedness of nature and human action. This period saw the emergence of environmental journalism as a new beat within the media industry—one that sought to interpret complex scientific findings for the general public and to hold industries and governments accountable (Hansen, 2010).

The 1970s also witnessed the institutionalization of environmental issues within political agendas, leading to the establishment of the United Nations Environment Programme (UNEP) in 1972 and the first Earth Summit in Stockholm the same year. These events introduced a global communicative dimension to environmental governance, as information exchange, public education, and media advocacy became recognized as key tools for fostering environmental awareness (Servaes, 2018). However, at that time, most communication efforts followed a *deficit model*—based on the assumption that environmental problems could be solved through the one-way transmission of scientific information from experts to citizens (Cox & Pezzullo, 2021). This instrumental approach overlooked the sociocultural and political factors shaping how people interpret environmental messages.

2.1.2 The Institutionalization of Environmental Communication (1990s)

By the early 1990s, environmental communication began to mature as an academic discipline distinct from environmental education or advocacy. The field consolidated around key scholarly venues such as the *Environmental Communication Interest Group* within the U.S. National Communication Association (NCA) and specialized journals like *Environmental Communication: A Journal of Nature and Culture* (founded in 2007). Theoretical advances during this period emphasized that communication does not simply reflect environmental realities but actively constructs them through discourse, framing, and representation (Hajer, 1995; Fairclough, 1995).

The 1992 United Nations Conference on Environment and Development (the Rio Earth Summit) provided a major boost to the field. It underscored the role of communication, participation, and education in promoting sustainable development, as articulated in *Agenda 21*. This recognition aligned with the growing influence of participatory development theory, which viewed communication as a dialogical process enabling communities to engage in decision-making rather than passive receivers of top-down messages (Freire, 1970; Servaes, 1999).

Thus, the 1990s marked the conceptual transition from communication *about* the environment to communication *for* the environment—signaling a paradigm shift from informing to engaging, from persuasion to dialogue, and from awareness to empowerment.

2.1.3 Globalization, Environmental Risk, and the Media (2000s–2010s)

In the early twenty-first century, environmental communication entered a new phase shaped by globalization, digitalization, and the rise of environmental risk as a defining feature of modernity. The works of Ulrich Beck (1992) on the *risk society* and Anthony Giddens (2009) on *climate politics* highlighted how communication processes mediate public understanding of global risks such as climate change, biodiversity loss, and pollution. The mass media became a primary arena for environmental meaning-making, where scientific, political, and cultural discourses intersect (Hansen & Machin, 2019).

Simultaneously, environmental communication scholars began examining how framing, metaphors, and narratives influence public perception of environmental issues. Terms such as *global warming*, *climate crisis*, and *ecological transition* reflect competing attempts to define the environmental agenda and mobilize public action (Nisbet, 2009). The 2000s also saw the emergence of digital environmental activism—social movements that use online platforms to shape discourse, organize collective action, and contest dominant narratives (Lester & Hutchins, 2013). These transformations marked a shift from institutional communication toward networked, participatory, and mediated environmental engagement.

2.1.4 Environmental Communication in the Global South

Although the field originated largely in Western academic contexts, environmental communication in the Global South evolved under different social, cultural, and political conditions. In regions such as Africa, Latin America, and the Arab world, environmental degradation is deeply entangled with issues of poverty, postcolonial dependency, and uneven development (Escobar, 1995). Communication practices therefore focus not only on awareness but also on empowerment, equity, and social justice.

In Algeria and the wider Maghreb, environmental discourse has historically been shaped by the legacy of colonial resource extraction and the centralization of state-controlled media. However, since the 1990s, environmental NGOs, local associations, and radio programs have played a growing role in fostering grassroots environmental awareness and community-based action (Rachwan, 2006; Ministry of Environment and Renewable Energies, 2023). Despite institutional constraints, these initiatives have contributed to the gradual emergence of a participatory environmental culture, particularly in rural and semi-urban areas.

At the same time, the Arab region's engagement with environmental communication has been largely policy-driven, influenced by regional institutions such as the League of Arab States and by international partnerships under frameworks like the *Arab Strategy for Sustainable Development (2030).* While scholarly literature in Arabic remains limited, the growing number of academic studies in recent years signals a rising awareness of the need to integrate communication within sustainability and climate governance frameworks (Al-Dulaimi, 2012).

2.1.5 The Digital Turn and the Age of Climate Communication (2015–Present)

The last decade has witnessed what scholars term the "digital turn" in environmental communication. With the proliferation of social media, podcasts, and citizen journalism, environmental issues have entered new communicative spaces where narratives, emotions, and data visualization play pivotal roles (Anderson, 2021). These digital platforms democratize environmental discourse but also introduce new challenges such as misinformation, selective exposure, and polarization.

Climate communication has become a distinct subfield within environmental communication, emphasizing the psychological, emotional, and social dimensions of how people perceive and respond to climate change. Research increasingly focuses on affective engagement, visual storytelling, and participatory design as tools for fostering behavioral change (Boykoff, 2019). In the Arab and African contexts, digital media have enabled young activists, journalists, and educators to bypass traditional communication hierarchies and to articulate localized forms of ecological consciousness.

In summary, the historical evolution of environmental communication reflects a broad epistemological transformation—from informational to participatory, from national to global, and from linear to networked. It is now recognized as an interdisciplinary field that unites communication theory, environmental sociology, cultural studies, and political ecology in addressing one of humanity's most urgent challenges: how to live sustainably within the limits of the planet.

2.2 Defining the Field: Concepts and Boundaries

The concept of environmental communication defies simple definition because it sits at the crossroads of multiple disciplines—communication studies, environmental sciences, sociology, political ecology, and cultural studies. Scholars have conceptualized it variously as a *process*, a *practice*, and a *field of inquiry*, each emphasizing distinct but interconnected dimensions of how humans construct, negotiate, and circulate meanings about the environment.

2.2.1 Environmental Communication as a Process

At its most basic level, environmental communication refers to the symbolic exchange of information and meanings related to environmental issues (Cox & Pezzullo, 2021). It encompasses all forms of communication—verbal, visual, institutional, or mediated—that influence the way individuals and societies perceive and act upon the natural world. This definition foregrounds the idea that "the environment" is not simply a physical entity but also a discursive and cultural construct shaped through language, images, narratives, and policy frames (Hajer, 1995; Dryzek, 2020).

Communication, in this sense, is not neutral: it mediates the power relations that determine whose voices are heard, whose knowledge counts, and which visions of sustainability prevail. Whether through news reporting, scientific communication, policy advocacy, or grassroots activism, environmental messages always operate within social fields of meaning and contestation. Thus, to study environmental communication is to study the politics of representation, persuasion, and participation surrounding ecological questions.

2.2.2 Environmental Communication as a Practice

From a practical standpoint, environmental communication also refers to the strategic use of communication tools and media to inform, educate, and mobilize audiences toward environmental protection and sustainable development. This applied perspective includes environmental journalism, public relations, risk communication, social marketing, and participatory education programs (Hansen & Machin, 2019; Lester & Hutchins, 2013).

Practitioners—journalists, educators, policymakers, and NGOs—rely on communication to shape public awareness, influence policy, and foster behavioral change. Yet, as Servaes (2018) notes, effective environmental communication requires more than technical dissemination; it demands a dialogical and culturally sensitive approach that recognizes local knowledge, traditions, and social contexts. The shift from "communicating to" audiences toward "communicating with" communities marks one of the defining transformations of the field.

2.2.3 Environmental Communication as an Academic Field

As an academic discipline, environmental communication emerged formally during the late 1980s and early 1990s. It was initially defined by Cox (1982, 2007) as "the pragmatic and constitutive vehicle for our understanding of the environment as well as our relationships with the natural world." This dual definition—pragmatic (communication as an instrument of advocacy and education) and constitutive (communication as a process that creates environmental meaning)—continues to guide much of the scholarly debate.

The field's boundaries remain dynamic and interdisciplinary. It draws theoretical insights from:

- **Rhetorical Studies**, focusing on persuasion, narrative, and moral reasoning;
- **Media and Cultural Studies**, exploring representation, imagery, and ideology;
- **Sociology and Political Ecology**, examining power, inequality, and governance; and
- **Communication for Development (C4D)**, emphasizing participation and empowerment in environmental contexts (Servaes, 2018; Manyozo, 2012).

This interdisciplinarity, while enriching, also presents conceptual challenges. Some scholars argue that environmental communication risks fragmentation, as it spans domains from media analysis to climate risk perception and advocacy research (Anderson, 2021). Others, however, view this hybridity as a strength that mirrors the complexity of environmental problems themselves, which demand integrative and multi-sectoral thinking (Dryzek, 2020).

2.2.4 Distinguishing Environmental Communication from Related Fields

It is essential to distinguish environmental communication from related but distinct fields:

1. **Environmental Education** aims primarily to transmit ecological knowledge and foster environmental literacy, often within formal or non-formal educational settings.
   In contrast, environmental communication extends beyond pedagogy to include public discourse, media framing, and social mobilization.

2. **Environmental Advocacy** focuses on persuasion and activism to achieve specific policy or behavioral goals.Environmental communication includes advocacy but also analyzes the underlying narratives, ideologies, and communicative strategies through which advocacy operates.
3. **Risk and Crisis Communication** addresses the dissemination of information during environmental hazards (e.g., oil spills, wildfires, pollution). While environmental communication intersects with this domain, it also incorporates broader cultural and political questions about how societies define and prioritize risks.

Hence, environmental communication operates at the interface of knowledge, power, and culture, serving as both a scientific inquiry and a civic practice that mediates humanity's evolving relationship with nature.

2.2.5 Toward an Operational Definition

Synthesizing these perspectives, this book defines environmental communication as:

"The multi-layered process through which individuals, institutions, and societies produce, circulate, and contest meanings about the environment, shaping both social practices and policy frameworks that guide collective responses to ecological change."

This definition underscores four central characteristics:

1. **Symbolic and cultural** – Environmental meanings are socially constructed through language and discourse.
2. **Political and ideological** – Communication processes reflect and reproduce relations of power, inclusion, and exclusion.
3. **Participatory and dialogical** – Sustainable change requires mutual engagement and shared understanding among diverse actors.
4. **Transformative and developmental** – Communication serves as a catalyst for environmental ethics, civic participation, and institutional reform.

In adopting this holistic perspective, the field transcends disciplinary silos to offer a comprehensive understanding of how environmental problems—and their potential solutions—are constructed, communicated, and contested across societies.

2.3 The Evolution of Theoretical Approaches

The intellectual trajectory of environmental communication mirrors the broader evolution of communication theory itself—from instrumental and linear paradigms of information transfer toward dialogical, critical, and ecological perspectives that view communication as an inherently social, cultural, and political process. This section traces these theoretical shifts, illustrating how each stage redefined the relationship between communication, society, and the environment.

2.3.1 The Instrumental and Informational Paradigm

The earliest conceptualizations of environmental communication were largely instrumental, rooted in the positivist logic of post–World War II modernization theory.

Communication was viewed as a neutral conduit for the dissemination of information designed to influence public understanding and behavior (Rogers, 1962). Environmental problems were perceived as technical challenges requiring expert knowledge and public compliance. Within this framework, communication served primarily to *inform* and *educate*—to bridge the so-called "knowledge gap" between scientists and citizens (Cox, 2007).

This "deficit model" presumed that environmental ignorance stemmed from a lack of information; thus, providing facts and data would naturally lead to pro-environmental behavior. However, by the late 1970s, researchers began to question the efficacy of this model, observing that awareness alone rarely translated into action (Hansen, 2010).

Critics argued that such an approach underestimated the roles of culture, ideology, and power in shaping environmental discourse (Carvalho, 2007).

In Algeria and much of the developing world, this instrumental logic still informs many governmental communication campaigns, where the environment is presented as a managerial or technical issue rather than a social and cultural one (Ministry of Environment and Renewable Energies, 2023). This underscores the enduring influence of the early paradigm and the need for a more participatory and reflexive approach.

2.3.2 The Persuasive and Advocacy Model

During the 1980s and 1990s, environmental communication evolved into a field of strategic advocacy aimed at mobilizing public opinion and influencing policy. The focus shifted from information dissemination to persuasion and agenda-setting, with communication seen as a tool for environmental campaigning, lobbying, and behavior change.

Rhetorical scholars such as Killingsworth and Palmer (1992) and Cox (1982, 2010) introduced environmental rhetoric as a means of analyzing how arguments, metaphors, and narratives construct public understandings of nature. The "green rhetoric" of this period emphasized the moral and emotional appeals used by activists and policymakers to motivate environmental concern.

Simultaneously, the agenda-setting theory (McCombs & Shaw, 1972) was applied to explore how media influence public priorities by framing certain issues as urgent or legitimate. Environmental coverage was found to fluctuate with political and economic cycles, often amplifying crises (such as oil spills or wildfires) while neglecting systemic causes (Anderson, 2021). This stage marked the beginning of a critical awareness that communication shapes not only attitudes but also the very structure of environmental knowledge and policy visibility.

2.3.3 The Participatory and Dialogical Turn

The 1990s witnessed a paradigm shift inspired by Paulo Freire's (1970) theory of dialogical communication and the emergence of Communication for Development (C4D) and Communication for Social Change (CFSC) frameworks (Servaes, 1999, 2018). These approaches reconceptualized communication as a two-way, horizontal process emphasizing participation, empowerment, and collective learning.

Applied to environmental contexts, this participatory paradigm called for inclusive communication strategies that integrate local knowledge systems, cultural values, and community voices into environmental decision-making. It also aligned with the concept of *environmental citizenship*, highlighting individuals' right to access, interpret, and influence environmental information (Manyozo, 2012).

The participatory model fundamentally challenged the technocratic assumptions of the deficit and advocacy paradigms. It argued that sustainable environmental change cannot be imposed externally but must be co-created through dialogue, negotiation, and shared responsibility.

This perspective found strong resonance in rural contexts, including Algeria, where community traditions such as *el-jema'a* (local council) and *touiza* (collective work) historically embodied participatory governance and resource sharing (Rachwan, 2006).

2.3.4 The Discursive and Critical Paradigm

The turn of the twenty-first century marked the discursive and critical turn in environmental communication. Influenced by the works of Fairclough (1995), Foucault (1972), and Hajer (1995), scholars began to view environmental issues as discursive constructions—that is, as outcomes of linguistic, institutional, and ideological processes that define what counts as "nature," "risk," or "sustainability."

This approach, often termed critical discourse analysis (CDA), examines how power operates through environmental narratives, metaphors, and representations. For instance, Hajer's (1995) concept of "storylines" demonstrates how policy actors use language to stabilize complex environmental debates and legitimize specific solutions. Similarly, Carvalho (2007) revealed how climate-change coverage in European media reflects broader ideological struggles over modernity, responsibility, and economic growth.

From this standpoint, communication is not simply about persuasion or participation—it is a form of power and governance.

It determines who can speak, which perspectives dominate, and which are silenced. This discursive orientation has been particularly influential in analyzing media representation of environmental issues in developing countries, where global narratives often overshadow local realities (Escobar, 1995; Dryzek, 2020).

2.3.5 The Ecological and Systems-Based Paradigm

More recently, scholars have proposed ecological models of communication that view communication itself as an ecosystem—a network of interdependent relationships between humans, media, technologies, and the natural world (Kalin, 2013; Anderson,

2021). This ecological turn emphasizes interconnectivity, feedback, and co-evolution, mirroring the principles of ecological systems.

In this paradigm, communication is both material and symbolic. Digital infrastructures, energy use, and technological consumption are recognized as integral components of the communication-environment nexus. The ecological paradigm thus extends beyond discourse to include the material dimensions of communication—the carbon footprint of media production, the environmental cost of data centers, and the sustainability of communication technologies (Hess, 2016).

For the Global South, adopting an ecological-systems perspective provides an opportunity to link environmental communication to broader debates on environmental justice, digital equity, and sustainable development.

It invites reflection on how local communities can use communication technologies responsibly to advance environmental goals while minimizing ecological impact.

2.3.6 A Synthesis: From Communication *About* to Communication *As* Environment

Across these paradigms, the evolution of environmental communication reveals a profound epistemological shift—from communication *about* the environment (as an object of discussion) to communication *as* environment (as a living system of relationships and meanings). This reflexive understanding acknowledges that communication is not external to the environment—it is part of it.

Language, media, and symbolic practices are themselves ecological forces shaping how societies interact with the biosphere. This perspective resonates strongly with indigenous and postcolonial worldviews, which reject the dualism between humans and nature and instead advocate for a communicative ethics grounded in interdependence and reciprocity (Al-Dulaimi, 2012; Rachwan, 2006).

Thus, the theoretical evolution of environmental communication represents not only an academic progression but also a paradigmatic transformation in human thought—from domination to dialogue, from control to coexistence, and from isolation to interconnectedness.

2.4 The Discursive and Sociocultural Dimensions of Environmental Communication

The study of environmental communication cannot be divorced from the linguistic, cultural, and ideological contexts through which environmental meanings are produced and circulated. If earlier paradigms emphasized communication as a tool for awareness or persuasion, the **discursive turn** reframed it as a *constitutive practice* that shapes how societies understand "nature," "development," and "sustainability."

This section explores how discourse, culture, and ideology interact to structure environmental communication, drawing upon critical traditions from linguistics, sociology, and cultural studies.

2.4.1 Discourse as the Construction of Environmental Reality

The foundational premise of **discourse theory** is that language does not merely reflect reality; it actively constructs it. According to Foucault (1972), discourse organizes knowledge by determining what can be said, who can speak, and what counts as truth within a given historical and institutional framework. In the context of the environment, discourses are systems of meaning that define nature not as an objective given but as a social and political construct (Hajer, 1995; Dryzek, 2020).

For example, the term "sustainable development," introduced by the Brundtland Report (1987), embodies a powerful discourse that reconciles economic growth with ecological responsibility. Yet, as Escobar (1995) and Fairclough (1995) argue, such discourses often conceal power asymmetries by naturalizing certain visions of progress while marginalizing alternative, local, or indigenous conceptions of living sustainably. Environmental communication thus involves an ongoing struggle over *definition* and *representation*—a contest between competing discourses that seek to establish legitimacy and authority.

In practical terms, these discursive constructions shape how environmental issues are framed in media narratives, government policies, and public debates.

They determine whether deforestation is represented as an economic necessity or an ecological crime, whether climate change is seen as a global emergency or a distant abstraction, and whether rural communities are portrayed as agents of sustainability or obstacles to modernization (Carvalho, 2007; Boykoff, 2019).

2.4.2 Critical Discourse Analysis and Power

Critical Discourse Analysis (CDA) provides a methodological and theoretical foundation for examining how language and communication enact power relations in environmental contexts. Building on Fairclough (1995) and van Dijk (2000), CDA posits that discourse functions ideologically by legitimizing specific social orders and environmental policies while marginalizing others.

For instance, media coverage of climate change often privileges voices of scientific experts or political elites, reproducing technocratic models of governance that exclude local or lay perspectives.

Similarly, corporate environmental campaigns—often framed under "green" or "sustainability" rhetoric—may serve to depoliticize environmental problems by shifting responsibility to individual consumers rather than systemic structures (Lester & Hutchins, 2013).

Through such mechanisms, discourse becomes a form of symbolic power (Bourdieu, 1991)—the ability to define what is thinkable, sayable, and actionable in environmental matters. CDA reveals that every communicative act, from a news article to a government slogan, contributes to the ideological reproduction or contestation of environmental meaning.

This critical approach has particular resonance in the Arab and Algerian contexts, where environmental communication often operates within tightly regulated media systems. Here, environmental discourse tends to reflect state priorities—such as national sovereignty, agricultural modernization, or energy policy—more than ecological pluralism.

Yet, alternative discourses persist through civil society networks, local radio, and academic circles, gradually widening the space for critical and participatory dialogue (Rachwan, 2006; Ministry of Environment and Renewable Energies, 2023).

2.4.3 The Cultural Mediation of Environmental Meaning

The sociocultural dimension of environmental communication acknowledges that meanings about nature are mediated by cultural values, collective memory, and symbolic practices. Culture determines not only how people perceive environmental change but also how they interpret responsibility, risk, and sustainability.

Hall's (1997) theory of *representation* suggests that meaning is produced through cultural codes that organize perception. In environmental communication, these codes include metaphors ("Mother Earth," "green revolution"), moral narratives (purity, stewardship, sacrifice), and visual imagery (forests, droughts, polluted rivers).

Such representations influence public emotion and cognition, shaping environmental engagement and policy support (Hansen & Machin, 2019).

Cultural contexts also define how communities interact with the natural world.

In many non-Western societies, including Algeria, traditional worldviews do not separate humans from nature but rather conceive of an integrated cosmology of coexistence. Concepts such as *el-baraka* (blessing) or *amane* (trust) embody a moral ecology that links human well-being with environmental harmony.

Integrating these indigenous perspectives into communication frameworks enriches environmental discourse with ethical depth and cultural legitimacy (Al-Dulaimi, 2012).

2.4.4 Ideology, Framing, and Environmental Hegemony

Beyond language and culture, ideology plays a central role in structuring environmental discourse. Following Gramsci's notion of *hegemony* (1971), dominant environmental narratives often reflect the interests of powerful actors—states, corporations, or international institutions—seeking to maintain control over the definition and management of nature.

Media studies have shown that the *framing* of environmental issues is rarely neutral.

Frames such as "economic development," "technological innovation," or "national security" can subtly steer public understanding away from questions of justice, equity, or sustainability (Nisbet, 2009). For instance, in fossil-fuel–dependent economies, media and political discourse frequently frame energy debates around national progress and sovereignty, while downplaying the environmental and social costs of extraction.

In Algeria, the discourse of *energy transition* illustrates this tension. While official communication highlights renewable energy as a symbol of modernization and international competitiveness, it often marginalizes public dialogue on local participation, ecological ethics, or social justice. The ideological dimension of environmental communication thus requires constant scrutiny—not only to reveal who speaks for the environment but also to interrogate who benefits from dominant narratives (Dryzek, 2020).

2.4.5 Intercultural and Postcolonial Perspectives

A comprehensive understanding of environmental communication must also account for intercultural and postcolonial dynamics.

As Escobar (1995) contends, the very concept of "development" emerged within a Western epistemology that positioned the Global South as both object and subject of intervention. Environmental communication in these contexts is often shaped by *discursive dependency*—the reliance on imported concepts, terminologies, and media frameworks that may not align with local realities.

Postcolonial communication scholars advocate for epistemic pluralism—the recognition of multiple ways of knowing and communicating about the environment (Servaes, 2018).

In Algeria, as in many postcolonial societies, reconciling scientific discourse with traditional ecological knowledge is a critical challenge.

Incorporating indigenous communication forms, oral storytelling, and community rituals into environmental education and advocacy can contribute to what Servaes calls *cultural sustainability*—the preservation of diversity in how environmental meanings are produced and shared.

2.4.6 Summary: Discourse, Culture, and Power in Environmental Communication

To summarize, the discursive and sociocultural dimensions of environmental communication reveal that:

1. **Environmental meanings are discursively constructed** through language, narrative, and representation rather than pre-given facts.
2. **Power operates through discourse**, shaping which environmental voices and perspectives are legitimized or marginalized.

3. **Culture mediates environmental perception**, embedding ecological values within moral, religious, and symbolic systems.
4. **Ideology organizes environmental hegemony**, aligning ecological narratives with political and economic interests.
5. **Postcolonial and intercultural approaches** challenge Western epistemologies, promoting pluralism and communicative justice.

Recognizing these dimensions is essential for developing communication strategies that are not only informative but also transformative—rooted in cultural meaning, critical reflection, and participatory ethics. In this way, environmental communication becomes both a mirror and an agent of societal change, capable of reimagining the human–nature relationship through discourse, dialogue, and shared cultural understanding.

2.5 Environmental Communication in the Arab and Algerian Contexts

2.5.1 Environmental Communication in the Arab World

In the Arab region, environmental communication has historically developed within the framework of state-centered modernization policies rather than grassroots environmental activism. The first regional efforts date back to the 1980s, when **the Arab League** and the Arab Organization for Administrative Development began producing guides and training materials to promote environmental awareness among youth and civil servants (Al-'Awadhi, 1992; Al-'Awadhi, 1996).

These initiatives introduced a discourse of environmental protection closely tied to national development and public administration, emphasizing moral responsibility and collective duty toward nature.

However, the communication model adopted during this period remained top-down and technocratic, focusing on transferring expert knowledge rather than facilitating community dialogue.

The absence of participatory media structures limited citizens' engagement in environmental governance. As a result, environmental awareness campaigns were largely confined to public institutions and formal education systems (Hanawi, 2004).

By the 2000s, environmental communication in Arab countries began to intersect with sustainable development policies, particularly after the 1992 Rio Earth Summit and the introduction of the *Arab Strategy for Sustainable Development (2002–2015).*

New regional media platforms, such as *Al-Bi'a wal-Tanmiyya* (Environment and Development), played a critical role in creating a transnational Arab environmental public sphere. Scholars like Al-Dulaimi (2012) highlighted the importance of combining Islamic ethical frameworks—such as stewardship (*khilāfa*) and balance (*mīzān*)—with modern ecological concepts to promote culturally grounded sustainability communication.

Recent years have also witnessed the rise of digital environmental activism in Arab contexts, where journalists, NGOs, and youth movements have used social media to disseminate ecological messages, challenge state narratives, and coordinate local actions. Yet, as Rachwan (2006) notes, these efforts face structural constraints, including restricted media freedom, limited access to environmental data, and dependency on international funding.

2.5.2 Environmental Communication in Algeria

In Algeria, environmental communication evolved within the broader postcolonial project of nation-building and modernization. The state's early environmental discourse emphasized reconstruction and resource management, particularly through large-scale programs like the *Green Dam* (*Barrage Vert*) initiated in the 1970s to combat desertification. Environmental communication during this era was primarily institutional and educational, focusing on technical awareness campaigns and radio programs that encouraged reforestation, hygiene, and agricultural productivity (Melha, 2000; Ministry of Environment and Renewable Energies, 2023).

Throughout the 1990s and 2000s, Algeria witnessed the gradual institutionalization of environmental governance.

The National Strategy for the Environment (2002) and subsequent national reports emphasized communication as a pillar of sustainable development.

Governmental communication bodies—such as the Ministry of Environment's outreach units—produced campaigns promoting waste management, biodiversity protection, and renewable energy.

However, these efforts were often criticized for adopting monologic communication styles, privileging expert discourse over public participation (Bessaoud, 2013).

Parallel to the official discourse, community-based environmental communication emerged through local associations, especially after the 1990 Law on Associations enabled the creation of environmental NGOs.

The region of Bouira provides a microcosm of this movement. Associations such as *Al-Amal for Nature Protection*, *Kawkabuna for Environmental Hygiene*, and *The Wilaya Association for Environment and Life* developed grassroots initiatives combining education, activism, and local media collaboration

These groups used cultural centers and radio programs to discuss waste management, forest conservation, and rural sanitation—embodying a participatory model of environmental communication.

This dual structure—institutional versus community communication—defines the Algerian experience.

The institutional level remains dominated by centralized planning, ministerial campaigns, and school-based education programs, while the community level demonstrates flexibility, cultural proximity, and experiential learning.

Together, they reflect a hybrid communication ecosystem where environmental messages are mediated between state authority and civic initiative.

2.5.3 Challenges and Prospects

Despite progress, environmental communication in Algeria faces persistent challenges:

1. **Institutional fragmentation** among ministries and local authorities weakens message coherence.
2. **Limited environmental journalism** results in superficial or event-driven coverage rather than investigative reporting.
3. **Low scientific literacy** among audiences hampers comprehension of environmental data.
4. **Insufficient integration of local knowledge**, such as traditional irrigation systems and communal forest management, reduces cultural resonance.

Nevertheless, opportunities abound. The expansion of digital media and youth-led environmental associations offers new pathways for participatory communication. Podcasts, Facebook pages, and citizen journalism projects now supplement traditional outlets, allowing environmental narratives to circulate beyond institutional boundaries.

Future progress will depend on establishing a multi-stakeholder communication model that bridges the gap between government institutions, scientific communities, local NGOs, and citizens. Such integration can transform environmental communication from a reactive tool into a proactive driver of sustainable rural development.

2.5.4 Summary

Environmental communication in the Arab and Algerian contexts illustrates a gradual shift from state-centered dissemination to community-centered dialogue.

While the former emphasizes technical control and modernization, the latter highlights empowerment, identity, and cultural meaning. Algeria's experience demonstrates that sustainable environmental governance requires both levels to converge—where official communication ensures policy legitimacy and grassroots networks ensure social relevance.

2.6 Chapter Summary

The theoretical exploration of environmental communication in this chapter has traced its evolution from an instrumental, information-centered model to a multidimensional, critical, and culturally embedded field of inquiry.

The chapter demonstrated that communication is not a mere vehicle for transferring environmental knowledge but a constitutive process that creates, circulates, and legitimizes meanings about nature, sustainability, and development.

Historically, the field developed through several overlapping paradigms.

The instrumental paradigm viewed communication as a tool for informing and persuading publics, while the advocacy model introduced rhetorical and strategic dimensions.

The participatory paradigm redefined communication as a dialogical and empowering process that integrates local knowledge and civic engagement.

Subsequently, the discursive and critical approaches—inspired by Foucault, Fairclough, and Hajer—revealed how language and ideology structure environmental governance. Finally, the ecological systems paradigm repositioned communication itself as an ecological process embedded within material and technological networks.

The chapter also highlighted the sociocultural and ideological dimensions of environmental discourse. It emphasized that cultural symbols, metaphors, and moral narratives mediate people's understanding of environmental change, while ideological frames shape who speaks for nature and whose perspectives are marginalized.

These insights call for a pluralistic, culturally grounded approach to environmental communication that recognizes the diversity of worldviews and experiences.

In the Arab and Algerian contexts, environmental communication has followed a unique trajectory—from state-centered awareness campaigns to community-based environmental activism.

The institutional discourse remains characterized by centralization and technical rationality, whereas grassroots associations and local media embody participatory and culturally resonant practices. Algeria thus exemplifies a hybrid communication model, in which institutional legitimacy and community engagement must converge to achieve sustainable outcomes.

Overall, this chapter situates environmental communication as both an academic field and a social practice—*a space where language, culture, and power converge to define humanity's relationship with the natural world.* It provides the theoretical foundation upon which the next chapter will build, exploring how communication interacts with sustainability and development paradigms, and how these relationships shape environmental governance and policy across diverse contexts.

Chapter 3 – Communication, Sustainability, and Development Paradigms

3.1 Conceptual Linkages between Communication and Sustainable Development

The relationship between communication and sustainable development is neither incidental nor secondary—it is structural and constitutive.

Communication serves as the cognitive, cultural, and political infrastructure that allows societies to define, negotiate, and operationalize the concept of sustainability.

Without communication, there is no shared vision of the future, no public deliberation on environmental priorities, and no collective capacity to act toward long-term ecological and social balance.

This section examines how communication has become central to the discourse and practice of sustainable development, tracing its conceptual foundations, its evolution across global policy frameworks, and its transformation into a participatory, multi-stakeholder process.

3.1.1 From Development to Sustainable Development: A Communicative Transition

The notion of *sustainable development* gained prominence following the 1987 Brundtland Report (*Our Common Future*), which defined it as "development that meets the needs of the present without compromising the ability of future generations to meet their own needs."

This definition inherently implies communication—it assumes dialogue across generations, sectors, and disciplines.

Initially, development communication in the postwar era was informed by modernization theory, which viewed communication as an instrument to transfer knowledge from the "developed" to the "developing" world (Rogers, 1962).

The focus was on diffusion of innovations—how to persuade farmers, workers, and citizens to adopt technologies and practices associated with industrial progress.

By the 1970s and 1980s, however, social scientists began to criticize this model for ignoring the structural inequalities and cultural diversity that shape development outcomes.

The idea of "communication for development" evolved into communication for sustainable development, integrating environmental, social, and ethical dimensions (Servaes, 1999; Manyozo, 2012).

Thus, sustainability emerged not merely as an economic or ecological condition but as a communicative construct—a shared narrative negotiated among multiple actors.

3.1.2 Communication as a Pillar of the Sustainable Development Goals (SDGs)

The United Nations' Sustainable Development Goals (SDGs) (2015–2030) provide a global framework where communication is not a separate goal but a *cross-cutting enabler*. Goals such as Quality Education (SDG 4), Climate Action (SDG 13), and Partnerships for the Goals (SDG 17) all hinge on effective communication strategies.

UNESCO and UNDP (2019) emphasize that sustainable development cannot succeed without *participatory communication*—the process through which communities articulate needs, deliberate options, and co-create solutions.

Environmental communication, in particular, translates complex issues such as climate change, biodiversity, and renewable energy into accessible narratives that foster understanding and mobilization (Cox & Pezzullo, 2021).

To illustrate this evolution, the following table compares the three historical stages of development communication, highlighting how the communicative logic shifted from transmission to participation and ultimately to governance.

Table 1. Evolution of Development Communication Paradigms

Stage	Dominant Paradigm	Historical Period	Core Communication Model	Main Characteristics	Implications for Sustainability
Stage I	*Modernization/ Diffusion Model*	1950s–1970s	Linear (Sender → Message → Receiver)	Top-down information flow; emphasis on behavioral change; reliance on experts and mass media	Promotes technological adoption but ignores cultural and ecological contexts
Stage II	*Dependency / Critical Model*	1970s–1980s	Dialogical and critical	Focus on structural inequalities; critiques global power asymmetry; emphasis on local agency	Introduces social justice and participation into development debates
Stage III	*Participatory/ Sustainability Model*	1990s–Present	Networked, multi-stakeholder	Emphasizes dialogue, empowerment, cultural diversity, and environmental ethics	Positions communication as a process of governance and shared responsibility

Sources: Rogers (1962); Freire (1970); Servaes (1999, 2018); UNDP (2019); Manyozo (2012).

3.1.3 Communication and the Three Pillars of Sustainability

Sustainability rests upon three interdependent pillars—environmental, economic, and social—each of which requires communication to function effectively:

1. **Environmental sustainability** depends on transparent communication about ecological risks, policy measures, and behavioral change. It transforms scientific knowledge into public understanding and collective action.
2. **Economic sustainability** relies on dialogue between producers, consumers, and regulators to ensure responsible production and consumption. Communicative practices such as corporate environmental reporting and green marketing are central to this domain.

3. **Social sustainability** is built through media representation, civic engagement, and cultural exchange. It involves communication that fosters inclusion, trust, and participation in environmental decision-making.

Thus, communication is the connective tissue that integrates the environmental, economic, and social dimensions into a coherent sustainability system.

3.1.4 Beyond Information: Toward Communicative Governance

Modern sustainability policies increasingly recognize that effective environmental governance is not a matter of information provision but of communicative participation. UNDP (2023) describes communication as a "fourth dimension of governance," alongside transparency, accountability, and participation.

Communicative governance emphasizes dialogue between institutions and citizens, transforming decision-making from a closed administrative process into a deliberative public arena.

In Algeria, for example, participatory radio programs and community workshops organized by local NGOs (such as *Association Al-Amal* and *Kawkabuna*) have begun to bridge the gap between citizens and environmental authorities.

This approach redefines sustainability as a relational process—a dynamic negotiation among diverse stakeholders. It treats communication not as a neutral medium but as a site of social co-creation, where environmental futures are collectively imagined and politically enacted.

3.1.5 Toward a Framework of Communicative Sustainability

Synthesizing these developments, we can conceptualize communicative sustainability as an integrative framework where:

- **Information** becomes *knowledge* through shared meaning-making.
- **Participation** transforms audiences into *agents* of change.
- **Governance** becomes *communication-based*, fostering trust and accountability.

This framework will guide the analyses in subsequent sections, where sustainability paradigms are examined in light of communication theory, policy practice, and the Algerian context.

3.2 Major Paradigms of Development Communication

The evolution of development communication has been shaped by successive theoretical paradigms reflecting broader shifts in the understanding of progress, participation, and sustainability.

Each paradigm embodies distinct assumptions about knowledge, power, and social change—and consequently about the role of communication in development processes.

From the technocratic optimism of modernization theory to the critical awareness of dependency thought and the empowerment focus of participatory approaches, these paradigms have defined how communication has been used (and sometimes misused) in the pursuit of development.

Understanding them is essential for contextualizing environmental communication within sustainable development frameworks.

3.2.1 The Modernization Paradigm: Communication as Transmission

The modernization paradigm, dominant from the 1950s through the 1970s, emerged from Western models of economic growth and social change.

It viewed development as a linear process of transformation from traditional to modern societies, driven by industrialization, urbanization, and technological innovation (Lerner, 1958; Rogers, 1962).

Communication, in this context, was conceptualized as a one-way transmission of modern values, ideas, and technologies from "developed" nations to "underdeveloped" ones.

The primary goal was to foster behavioral change among target populations—farmers, villagers, or workers—so they would adopt innovations perceived as markers of progress.

The mass media—radio, press, cinema—were considered powerful instruments for modernization. Early communication programs in Algeria during the 1960s and 1970s mirrored this paradigm: agricultural radio programs, literacy campaigns, and industrial propaganda were designed to integrate rural populations into the national development project (Melha, 2000).

However, modernization theory faced strong criticism for its ethnocentrism and technocratic bias. It ignored cultural diversity, power inequalities, and local knowledge systems. It assumed that information alone could produce development—a notion that proved empirically weak and ideologically problematic (Servaes, 1999).

3.2.2 The Dependency Paradigm: Communication as Resistance and Awareness

In the 1970s, Latin American scholars such as Paulo Freire (1970), Luis Ramiro Beltrán (1976), and Orlando Fals Borda (1985) proposed a dependency paradigm that directly challenged modernization theory.

Drawing from Marxist and critical sociology, this approach argued that underdevelopment was not a lack of modernization but a product of structural dependency—the unequal economic and informational relations between the Global North and South.

From this perspective, communication was not a neutral transmitter of progress but a site of ideological domination.

Mass media in developing countries were often controlled by elites, reproducing Western models of consumption and progress while silencing local voices (Beltrán, 1976).

Freire's notion of *dialogical communication* provided a radical alternative: communication should not seek to persuade but to conscientize—to raise critical awareness (*conscientização*) among marginalized groups so they can reflect upon and transform their reality.

In the Algerian and Arab contexts, this perspective found resonance in educational and rural development programs that emphasized local participation and cultural identity.

For instance, the *Green Dam* initiative (Barrage Vert) and agricultural extension programs began to incorporate elements of awareness and civic engagement, moving beyond pure instruction to stimulate discussion and feedback (Bessaoud, 2013).

3.2.3 The Participatory Paradigm: Communication as Empowerment

The participatory paradigm, which gained global recognition in the 1990s, built upon the critiques of dependency theorists and incorporated postmodern, feminist, and postcolonial insights.

It redefined communication as a horizontal process grounded in dialogue, cultural sensitivity, and community ownership (Servaes, 2018; Manyozo, 2012).

Participation was no longer seen as a means but as an end in itself—a manifestation of democratic development.

Within this framework, sustainable change depends on the inclusion of local actors in decision-making, problem definition, and evaluation.

In Algeria, participatory communication has been increasingly adopted in environmental and rural programs led by the Ministry of Environment, local radio stations, and associations such as Al-Amal and Kawkabuna.

These initiatives involve citizens directly in problem identification—whether reforestation, waste management, or water conservation—transforming them from passive recipients into active co-creators of solutions (Ministry of Environment, 2023).

Participatory communication thus represents both a normative ideal and a practical methodology.

It recognizes the cultural and political dimensions of development and integrates indigenous knowledge, collective decision-making, and gender inclusion into sustainability discourse.

3.2.4 Comparative Synthesis of Development Communication Paradigms

To summarize the evolution and implications of these paradigms, Table 2 presents a comparative overview highlighting their theoretical foundations, communicative logics, and relevance for environmental and sustainable development.

Table 2. Comparative Overview of Development Communication Paradigms

Dimension	Modernization Paradigm	Dependency Paradigm	Participatory Paradigm
Historical Context	1950s–1970s (Postwar industrialization and Cold War)	1970s–1980s (Postcolonial critique and Latin American thought)	1990s–Present (Globalization, democracy, sustainability)
Intellectual Roots	Structural functionalism, behavioral psychology	Marxism, critical sociology, liberation pedagogy	Constructivism, cultural studies, postcolonial theory
View of Development	Linear progression from "traditional" to "modern"	Structural dependence and unequal exchange	Locally defined, inclusive, and culturally embedded process
Communication Model	Linear and persuasive (sender → receiver)	Dialogical and critical	Horizontal, networked, participatory
Role of the Audience	Passive recipients of knowledge	Subjects of conscientization	Co-creators and decision-makers
Media Tools	Mass media, campaigns, propaganda	Alternative media, community press, education	Community radio, participatory video, social media
Examples in Algeria	Early radio literacy campaigns; agricultural extension under state control (1960s–1970s)	Grassroots environmental education; rural cooperatives (1980s–1990s)	Local NGOs, environmental associations, participatory workshops (2000s–present)
Limitations	Technocratic and ethnocentric; ignores culture	Over-politicized; limited institutional application	Resource-intensive; depends on political openness
Contribution to Sustainability	Promotes awareness of technology	Promotes critical awareness of inequity	Promotes empowerment and collaborative governance

Sources: Freire (1970); Beltrán (1976); Servaes (1999, 2018); Manyozo (2012); Bessaoud (2013); Ministry of Environment and Renewable Energies (2023).

3.2.5 Toward Integrative Communication for Sustainable Development

In practice, no paradigm operates in isolation. Effective environmental and sustainability communication requires hybrid models that combine the efficiency of modernization (information systems), the critical awareness of dependency theory (equity and justice), and the inclusiveness of participatory approaches (dialogue and agency).

In Algeria, this synthesis is increasingly visible in programs that blend top-down policy communication with bottom-up citizen participation—for instance, collaborative forest monitoring, waste recycling initiatives, and eco-education projects linking schools and municipalities.

Such integrative frameworks align with the idea of communicative sustainability, where the exchange of information is not merely instrumental but constitutive of governance, identity, and collective action. In this sense, communication becomes the *ecological infrastructure* of sustainable development.

3.3 Sustainability as a Communicative Process

The shift from "development" to "sustainable development" has redefined communication as more than an instrument of information—it has become the *medium through which sustainability itself is enacted*. In contemporary theory, sustainability is not only an ecological state or a policy framework but also a discursive and relational process: a continual negotiation among actors, institutions, and publics about how to live, consume, and govern within ecological limits.

3.3.1 Sustainability as a Discourse

Scholars such as Dryzek (2020) and Hajer (1995) describe sustainability as a *master discourse* that organizes environmental knowledge, policy, and ethics across global and local contexts. It functions as a "floating signifier" that can be filled with diverse meanings—ranging from "green growth" to "degrowth," from "eco-modernization" to "environmental justice."

This flexibility makes sustainability both powerful and ambiguous. It enables broad consensus while concealing ideological conflicts.

For example, governments may frame sustainability in terms of technological innovation and market efficiency, whereas civil society organizations may emphasize justice, inclusion, and cultural identity (Carvalho, 2007).

Hence, understanding sustainability requires critical analysis of how it is communicated—whose voices define it, whose values it reflects, and whose interests it serves. This brings us to the concept of communicative sustainability, which emphasizes dialogue, participation, and cultural translation as preconditions for ecological transformation.

3.3.2 The Role of Communication in Building Sustainable Societies

Communication sustains societies in both symbolic and material ways.

It enables individuals and communities to:

1. **Interpret** environmental problems (through media, education, and cultural narratives);
2. **Coordinate** actions and responsibilities (through deliberation, policy-making, and cooperation);
3. **Transform** values and behaviors (through ethical discourse and participatory engagement).

This triadic role—*interpretive, coordinative, transformative*—makes communication the invisible infrastructure of sustainability. As Servaes (2018) explains, sustainable development is "as much about communication as it is about economics or ecology," because it requires consensus, inclusion, and mutual learning.

In Algeria, for instance, environmental communication programs associated with rural development—such as waste separation projects in Bouira and reforestation drives in Blida—operate as communicative processes where local narratives, governmental discourse, and technical expertise intersect (Ministry of Environment and Renewable Energies, 2023).

3.3.3 The Ethical and Cultural Dimensions of Communicative Sustainability

Sustainability also carries ethical and cultural implications.

It is anchored in moral discourses that define what is "right" and "responsible" toward nature and future generations. The ethics of sustainability involve values such as solidarity, equity, and care—all of which are communicatively maintained through storytelling, education, and media representation (Pezzullo & Cox, 2021).

Cultural contexts deeply shape how these values are expressed.

In Arab and Algerian societies, for instance, religious and cultural narratives often associate environmental care with faith and moral duty. Concepts such as *khilāfa* (stewardship) and *amāna* (trust) position humans as moral custodians of the Earth (Al-Dulaimi, 2012). Integrating such culturally embedded ethics into environmental communication enhances its resonance and legitimacy.

3.3.4 Communication and Systemic Sustainability

From a systems perspective, communication functions as an ecological process—it circulates information, energy, and meaning within social and environmental networks (Kalin, 2013). Just as ecosystems rely on feedback loops for stability, sustainable societies depend on communicative feedback between citizens, institutions, and nature.

Failures in communication—such as misinformation, lack of transparency, or exclusion—can produce ecological and social instability. Conversely, strong communicative feedback mechanisms (like community monitoring, participatory reporting, and local dialogue platforms) enhance system resilience.

This systems-based approach aligns with the ecological modernization framework, which views communication as both a driver and a reflection of societal adaptation to environmental limits (Mol & Spaargaren, 2000).

3.3.5 Integrating Environmental, Social, and Communicative Dimensions

To understand how communication weaves through the fabric of sustainability, the following table summarizes its intersections across the environmental, social, and communicative pillars of sustainable development.

Table 3. Intersections between Environmental, Social, and Communicative Dimensions of Sustainability

Dimension	Core Objectives	Key Communicative Functions	Examples (Global / Algerian Context)
Environmental	Protect ecosystems, manage resources, mitigate climate change	Informing and educating about risks, promoting behavioral change, facilitating scientific translation	Awareness campaigns on forest fires (APS, 2022); school-based recycling programs in Bouira
Social	Promote equity, inclusion, participation, and cultural continuity	Enabling dialogue, building consensus, empowering marginalized groups	Community workshops on rural waste management (Kawkabuna Association); local women's cooperatives in Blida
Communicative	Ensure transparency, access to information, and public engagement	Co-creation of narratives, participatory governance, intersectoral coordination	Public environmental consultations; citizen journalism on climate impacts (Algerian online media)

Sources: Servaes (2018); Pezzullo & Cox (2021); Ministry of Environment and Renewable Energies (2023).

3.3.6 From Symbolic Awareness to Practical Transformation

Ultimately, communication transforms sustainability from a rhetorical ideal into a social practice. Awareness alone is insufficient; it must be followed by deliberation, collaboration, and behavioral change. The transition from symbolic discourse to practical transformation occurs through four communicative stages:

1. **Recognition** – Identifying environmental challenges (media, education).
2. **Deliberation** – Discussing solutions among stakeholders (forums, public hearings).
3. **Coordination** – Implementing shared actions (policies, community initiatives).

4. **Reflection** – Evaluating and learning collectively (feedback and adaptation).

Each stage reinforces the other in a cyclical process that mirrors natural ecosystems. This reflexive communication loop embodies the essence of sustainable societies—ones that learn, adapt, and evolve through communication.

3.3.7 Toward a Theory of Communicative Sustainability

Bringing together discursive, ethical, and systemic insights, the theory of communicative sustainability rests on three pillars:

- **Interdependence:** Humans, institutions, and environments are communicatively co-constituted.
- **Plurality:** Multiple worldviews and knowledge systems must coexist and interact dialogically.
- **Responsiveness:** Sustainability depends on open feedback loops, ethical reflection, and adaptive governance.

This theory does not replace economic or ecological models of sustainability but complements them, positioning communication as their *enabling condition*. For societies such as Algeria—where modernization, cultural heritage, and ecological vulnerability intersect—this perspective provides a pathway to harmonize development with environmental ethics.

3.4 Case Illustrations: Global and Algerian Experiences

3.4.1 Global Experiences in Communicative Sustainability

Across the world, environmental communication has evolved from awareness campaigns to multi-stakeholder participatory systems integrated within governance structures. The following global cases illustrate how communication is being used as a driver of sustainability rather than a supplementary tool.

1. **UNDP and Participatory Communication for Sustainable Development**
   The United Nations Development Programme (UNDP) has institutionalized communication as a fundamental pillar of sustainable governance.

Its *Guidelines on Communication for Sustainable Development (C4D)* (2019) emphasize the role of participatory dialogue in addressing poverty, inequality, and environmental degradation.

In countries such as Nepal, Philippines, and Costa Rica, UNDP-led initiatives have demonstrated how community radio, local storytelling, and public consultation platforms can transform environmental projects into socially owned processes. For instance, in Nepal's *Community Forestry Programme*, participatory radio programs enabled villagers to negotiate forest use collectively, improving both ecological outcomes and gender inclusion (FAO, 2020).

2. **UNESCO's Environmental Education Networks**
   UNESCO has promoted Education for Sustainable Development (ESD) through regional networks linking schools, NGOs, and local media.

Communication in this framework serves both educational and cultural functions—it builds shared ecological literacy while respecting local traditions. Projects in Morocco and Tunisia have shown how integrating environmental communication into curricula increases youth engagement and fosters intergenerational dialogue about sustainability (UNESCO, 2021).

3. **European Union and the Digital Transition**
4. The European Green Deal (EU, 2020) incorporates communication into environmental governance through transparency portals, open data initiatives, and participatory digital consultations. These mechanisms allow citizens to co-design policy frameworks and monitor progress toward carbon neutrality. The model underscores the concept of digital environmental citizenship, where technology enhances accountability and community learning (Dryzek, 2020).
5. **African Regional Experiences**
   In Kenya, community radio programs such as *Radio Mang'elete* have played a key role in mobilizing women's groups for water conservation and tree planting (Manyozo, 2012).
   Similarly, in Ghana, the *Eco-Schools Network* links teachers, journalists, and NGOs in participatory campaigns on waste reduction and renewable energy.

These experiences highlight how communication nurtures *collective ecological agency*—a shared sense of ownership over environmental futures.

3.4.2 The Algerian Experience: Institutional and Community Models

In Algeria, environmental communication has progressed through two interrelated models: the institutional (top-down) model driven by state policies, and the community (bottom-up) model led by NGOs, local media, and citizen initiatives.

A. Institutional Communication: Centralized Awareness and Policy Framing

The Ministry of Environment and Renewable Energies has played a key role in shaping environmental discourse since the 1990s.

Its communication strategy combines traditional awareness campaigns with policy-oriented dissemination.

Major initiatives include:

- The National Environmental Strategy (2002), which introduced nationwide campaigns on waste management and biodiversity conservation.
- The National Plan for Sustainable Development (PNDD, 2019), which integrated communication tools into climate adaptation projects.
- The creation of "Environment Weeks" and school-based competitions promoting eco-literacy.

According to official reports (Ministry of Environment, 2023), over 1,200 local awareness campaigns and 150 media partnerships were recorded between 2018 and 2022. The primary communication channels included television, public radio, and school education programs.

However, institutional communication often remains monologic and event-based, focusing on information transfer rather than participatory engagement.

The public's role is generally limited to reception, not deliberation—a challenge identified by Algerian communication scholars such as Rachwan (2006).

B. Community Communication: Grassroots Activism and Local Media

In contrast, community-based environmental communication in Algeria reflects a participatory and culturally embedded model.

Since the 1990 Law on Associations, NGOs and local groups have become active agents in environmental education and advocacy.

Notable examples include:

- **Association Al-Amal for Nature Protection (Bouira):**
  Conducts awareness campaigns on forest preservation, reforestation, and rural hygiene. Collaborates with schools and local radio (*Radio Soummam*) to disseminate environmental messages in Arabic, Tamazight, and French.
- **Kawkabuna Association for Environmental Hygiene (Bouira):**
  Focuses on waste sorting and urban cleanliness through participatory workshops and social media campaigns. Its "Clean Neighborhood" initiative (2019–2022) reached more than 5,000 residents.
- **El-Watan Radio Environmental Program:**
  A weekly segment that hosts journalists, farmers, and environmental officers to discuss local issues such as desertification, water scarcity, and sustainable agriculture.

These initiatives demonstrate that grassroots communication ensures cultural relevance and long-term behavioral change. Unlike institutional campaigns, community programs rely on interpersonal dialogue, shared experience, and trust.

Table 4. Comparison of Global and Algerian Environmental Communication Models

Aspect	Global Models (UNDP, UNESCO, EU)	Algerian Models (Institutional & Community)
Main Objective	Governance integration and participatory policy-making	Environmental awareness and behavioral change
Communication Approach	Multi-stakeholder dialogue, deliberation, co-design	Predominantly top-down in institutional model; participatory in community model
Tools Used	Digital platforms, participatory media, community radio, open data	Television, radio, schools, workshops, local NGOs, social media
Target Audience	Citizens, policymakers, civil society, youth networks	Rural populations, students, local communities
Degree of Participation	High (co-decision and feedback mechanisms)	Medium to low in official programs; high in local associations
Key Strengths	Inclusive governance, transparency, networked learning	Cultural proximity, grassroots mobilization, bilingual communication
Challenges	Digital inequality, sustainability of funding	Centralization, lack of coordination, limited journalistic depth
Notable Results	Institutionalized participatory communication frameworks	Growing NGO-led ecological literacy and civic engagement

Sources: UNDP (2019); UNESCO (2021); EU Green Deal (2020); Ministry of Environment and Renewable Energies (2023); Rachwan (2006).

3.4.3 Data-Driven Insights from the Algerian Context

Recent data illustrate Algeria's progress and remaining gaps in communicative sustainability:

- **Public Awareness Levels:**

  According to the Ministry of Environment (2023), 65% of citizens express concern about environmental issues, but only 28% participate in collective actions such as tree planting or waste collection.
  This reflects a gap between *awareness* and *engagement*—a classic communication challenge.
- **Media Representation:**
  A 2022 analysis of *APS* (Algerie Presse Service) environmental coverage found that 72% of reports were informational (announcing events or policies), while only 14% included public or expert debate (Bessaoud, 2013). This suggests a predominance of descriptive rather than deliberative discourse.
- **NGO Participation:**
  Over 500 environmental NGOs are officially registered in Algeria (UNEP, 2023), but only 15–20% maintain continuous media engagement or institutional partnerships.

These indicators underscore the need for integrated communicative governance, where public institutions, journalists, and civil society collaborate systematically.

3.4.4 Comparative Analysis: Lessons and Opportunities

The comparison between global and Algerian experiences yields several analytical insights:

1. **Global models prioritize deliberation and feedback**, while Algerian models emphasize awareness and compliance.
2. **Institutional fragmentation** in Algeria limits message coherence across ministries and regions.
3. **Local NGOs act as mediators**, translating national policies into culturally meaningful narratives.
4. **Digital tools remain underused**, despite growing internet penetration and youth online presence.

5. **Cultural embeddedness** offers Algeria a unique advantage: ecological narratives can draw upon spiritual, communal, and linguistic diversity.

The challenge is therefore not to import global models wholesale but to hybridize them—combining institutional resources with grassroots creativity to form a *localized ecosystem of environmental communication.*

3.4.5 Toward an Algerian Framework for Communicative Sustainability

Building on the lessons from both global and domestic contexts, Algeria can move toward a national model of communicative sustainability that integrates:

- **Institutional legitimacy** (policy coordination, data transparency),
- **Community participation** (dialogue and cultural resonance), and
- **Digital inclusion** (accessible platforms for civic engagement).

Such a framework would transform communication from a reactive awareness tool into a strategic instrument of governance and empowerment—a living infrastructure connecting policy, culture, and public action.

3.5 Toward a Model of Communicative Sustainability in Algeria

The preceding sections have shown that sustainable development cannot exist without sustainable communication—a process through which institutions, media, and citizens continuously exchange information, values, and responsibilities concerning environmental action.

In Algeria, where institutional environmental discourse often remains centralized and informational, the challenge is to construct a hybrid, participatory, and culturally grounded model of communication capable of translating sustainability from policy rhetoric into social practice.

3.5.1 The Rationale for an Algerian Model

Algeria presents a unique communicative environment. It combines:

- **A strong state apparatus** that manages environmental policy and resource governance;
- **Vibrant local communities** rooted in traditions of cooperation (*touiza*) and collective work;
- **Expanding digital ecosystems**, especially among youth, offering unprecedented opportunities for participatory engagement.

However, these components often operate in isolation. Institutional messages rarely reach rural communities effectively, while local initiatives often lack national visibility or coordination.

A model of communicative sustainability must therefore bridge these gaps through multi-level integration and two-way feedback mechanisms connecting state, society, and media.

3.5.2 Principles Guiding the Model

The proposed model rests upon five guiding principles that reflect both global best practices and Algeria's sociocultural realities:

1. **Transparency** – Ensuring open access to environmental data and decision-making processes.
2. **Participation** – Involving citizens, associations, and local leaders in all communication stages.
3. **Cultural Relevance** – Embedding messages within local languages, symbols, and narratives.
4. **Reflexivity** – Establishing continuous feedback loops for learning and policy adaptation.
5. **Integration** – Coordinating institutional, community, and digital communication channels.

These principles correspond to the ethical triad of *accountability, inclusion,* and *empowerment*, positioning communication as the foundation of sustainable governance.

3.5.3 Components of the Communicative Sustainability Model

The model is structured around three interconnected levels—institutional, community, and digital—each with specific functions and interaction mechanisms.

Table 5. A Model of Communicative Sustainability for Algeria

Level	Main Actors	Core Communicative Functions	Key Tools and Practices	Expected Outcomes
Institutional Level	Ministry of Environment, Local Authorities, Public Media	Policy dissemination, inter-sectoral coordination, environmental reporting	Press briefings, government portals, radio/TV programs, parliamentary hearings	Coherent national narrative; transparent policy communication; improved inter-ministerial collaboration
Community Level	Local NGOs, schools, religious leaders, farmers' associations	Dialogue, cultural mediation, local knowledge exchange	Workshops, theatre forums, participatory radio, festivals, storytelling	Enhanced public awareness, cultural ownership of sustainability, inclusive dialogue
Digital Level	Journalists, youth networks, influencers, universities	Networked participation, real-time feedback, data visualization	Social media campaigns, open data dashboards, citizen reporting apps	Broader engagement, faster information flow, civic monitoring and accountability

Sources: Adapted from Servaes (2018); UNDP (2019); Ministry of Environment and Renewable Energies (2023).

3.5.4 Operational Mechanisms: From Information to Interaction

For this model to function, communication must evolve through four operational mechanisms that translate theory into practice:

1. **Information Flow:**
   Institutional actors provide transparent, timely data on environmental initiatives (e.g., waste management, renewable energy, biodiversity).
2. **Dialogue and Consultation:**

Citizens and associations engage in discussions via public hearings, radio debates, and digital platforms to deliberate priorities.

3. **Co-creation:**
   Joint projects between ministries, local groups, and journalists ensure co-production of messages and actions—such as youth eco-clubs or school gardens.
4. **Feedback and Evaluation:**
   Periodic citizen surveys and digital reporting tools allow real-time evaluation of environmental campaigns and adaptation of policies.

These mechanisms transform communication from a vertical process of dissemination into a horizontal and reflexive ecosystem that strengthens collective learning and governance.

3.5.5 Cultural Anchoring of the Model

Sustainability communication in Algeria must not only be technically sound but also culturally resonant.

The use of local idioms, proverbs, and spiritual values can strengthen message legitimacy.

For example, campaigns that link environmental care to *amāna* (trust) and *khilāfa* (stewardship) align sustainability with Islamic ethical frameworks, while storytelling in Amazigh and local dialects reinforces inclusivity.

By acknowledging cultural diversity, environmental communication becomes a form of cultural empowerment—a dialogue between modern policy and ancestral wisdom.

3.5.6 Integration with Global Sustainability Frameworks

The proposed Algerian model aligns with international frameworks such as the UN Sustainable Development Goals (SDGs), especially:

- **SDG 13** (Climate Action),
- **SDG 16** (Peace, Justice, and Strong Institutions), and
- **SDG 17** (Partnerships for the Goals).

It also complements regional strategies like the Arab Strategy for Sustainable Development (2030), positioning Algeria as a potential leader in integrating communication within environmental governance across the Maghreb.

3.5.7 Challenges and Implementation Pathways

Despite its promise, the success of communicative sustainability in Algeria depends on addressing several structural and logistical challenges:

- **Institutional inertia:** Bureaucratic silos hinder cross-sector collaboration.
- **Limited media training:** Journalists often lack specialization in environmental reporting.
- **Digital divide:** Rural areas face connectivity constraints limiting online participation.
- **Funding sustainability:** Many local NGOs rely on short-term donor support.

To overcome these challenges, three strategies are essential:

1. **Capacity building** through training programs for journalists and civil servants in environmental communication;
2. **Partnership frameworks** between government, academia, and civil society to co-produce communication content;
3. **Digital inclusion policies** that ensure equitable access to information and participation.

3.5.8 Synthesis: Communication as Governance

In conclusion, the Model of Communicative Sustainability for Algeria envisions communication not as an accessory but as a core mechanism of governance.

It integrates environmental management with participatory democracy and digital innovation.

Through continuous interaction among institutions, communities, and networks, communication becomes the *social metabolism* of sustainability—a process through which knowledge, values, and actions circulate to maintain ecological and social equilibrium.

This model will serve as the conceptual bridge toward the following chapters, which will operationalize these insights in empirical and policy contexts, demonstrating how communicative governance can foster genuine sustainable transformation in Algeria and beyond.

Chapter 4 – From Theory to Practice – Applying Communicative Sustainability in Algeria's Environmental Governance

4.1 Introduction

Building upon the theoretical foundations established in previous chapters, this section operationalizes the concept of *communicative sustainability* within the Algerian context.
While earlier discussions emphasized the interplay between communication, culture, and environmental policy, this chapter explores how these dynamics manifest in the governance of environmental programs, focusing on the mechanisms through which information, participation, and accountability shape ecological outcomes.

The aim is twofold:

1. To evaluate how communication functions as a *governance instrument* in Algeria's environmental policy framework;
2. To propose concrete pathways for strengthening transparency, citizen engagement, and digital participation.

4.2 Conceptual Premise: Governance as Communication

Environmental governance today transcends administrative regulation—it operates as a communicative process of coordination, negotiation, and public accountability.

According to Dryzek (2020) and Meadowcroft (2011), effective governance requires not only laws and institutions but also channels for dialogue among policymakers, citizens, and knowledge producers.

In Algeria, environmental governance is anchored in a multi-tiered system involving:

- **The Ministry of Environment and Renewable Energies**,
- **Local environmental directorates (wilayas)**,
- **Public agencies** such as *Agence Nationale des Déchets (AND)* and *Conservation des Forêts*,
- **Civil society actors** including environmental NGOs, and

- **Media organizations**, both public and private.

However, as noted by UNDP (2019), the flow of information between these actors often remains fragmented, leading to policy duplication, delayed responses, and limited citizen engagement.

The communicative sustainability model, therefore, proposes viewing governance as a networked dialogue system, where each institutional node (state, media, civil society) contributes to a shared ecology of information and accountability.

4.3 Institutional Framework for Environmental Communication in Algeria

4.3.1 National Policies and Strategies

Algeria has adopted several strategic frameworks that integrate communication within environmental policy

Policy / Strategy	Year of Adoption	Key Communication Components	Implementing Institutions
National Environmental Strategy	2002	Awareness campaigns on waste, water, and biodiversity	Ministry of Environment
National Plan for Sustainable Development (PNDD)	2019	Integration of media partnerships and citizen participation	Inter-ministerial Committee
National Climate Plan	2020	Dissemination of adaptation measures, regional consultation forums	Ministry of Energy Transition and Environment
Green Algeria Initiative	2022	Public communication on renewable energy and circular economy	Ministry of Renewable Energies

Sources: Ministry of Environment (2023); UNDP (2022).

These strategies mark an institutional recognition that communication is essential for achieving sustainability goals.

Yet, implementation remains uneven due to limited inter-sectoral coordination and insufficient professionalization of environmental journalism.

4.4 Media and Environmental Discourse in Algeria

4.4.1 Quantitative Overview of Media Coverage

An analysis of Algerie Presse Service (APS) publications between 2020 and 2023 reveals the following distribution of environmental coverage:

Theme	Percentage of Coverage	Dominant Tone	Example of Headline
Climate Change and Wildfires	35%	Informational / Descriptive	"Les incendies ravagent 10.000 hectares dans l'Est du pays"
Waste Management and Recycling	28%	Normative / Policy-driven	"Vers une économie circulaire en Algérie"
Renewable Energy and Solar Projects	22%	Promotional / Optimistic	"L'Algérie investit dans le solaire pour diversifier son économie"
Water and Drought Issues	15%	Alarmist / Problem-oriented	"La rareté de l'eau menace les zones rurales"

Sources: APS Archive (2020–2023); Author's content analysis.

While coverage volume has increased, qualitative analysis shows a predominance of event-driven and top-down narratives.

Few reports include citizen perspectives or investigative depth, confirming earlier findings by Bessaoud (2013) and Rachwan (2006).

4.4.2 The Role of Local and Digital Media

Local radio stations (e.g., *Radio Soummam, Radio Sétif*) have proven more effective in fostering two-way environmental dialogue. Social media platforms such as Facebook and YouTube have also emerged as alternative public spheres, where environmental activists and NGOs like *Kawkabuna* or *Al-Amal Bouira* disseminate local success stories, mobilize volunteers, and share real-time environmental alerts.

This evolution signals a transition from centralized broadcasting to participatory networking, yet it also raises issues of information reliability and data fragmentation, especially when digital content lacks scientific verification.

4.5 Multi-Actor Dynamics: Communication Flows in Practice

4.5.1 Mapping Communicative Relations

To visualize the complexity of Algeria's environmental governance, Table 6 outlines the main communicative relations among institutional and societal actors.

Table 6. Communication Dynamics in Algeria's Environmental Governance

Actor	Primary Communication Partners	Type of Interaction	Communication Channel	Challenges Identified
Ministry of Environment	Media, NGOs, Schools, Municipalities	Top-down dissemination	Press releases, TV, official websites	Limited feedback and evaluation mechanisms
Local Authorities	Citizens, NGOs, Public Media	Consultative / Informational	Public meetings, local radio	Weak documentation and follow-up
NGOs and Civil Associations	Citizens, Media, Schools	Participatory / Awareness-building	Social media, workshops	Funding instability, lack of coordination
Journalists and Broadcasters	Government, Experts, Public	Agenda-setting / Framing	News reports, debates, field coverage	Lack of specialization in environmental reporting
Citizens and Youth Groups	Local authorities, NGOs, Online communities	Interactive / Advocacy	Social media platforms	Limited access to institutional data

This mapping highlights the fragmentation and asymmetry of communicative exchanges: while institutional actors dominate message creation, civil society remains central to message circulation and adaptation.

4.6 Evaluation of Current Communication Practices

According to a UNEP (2023) assessment, Algeria's environmental communication exhibits:

- High awareness potential (strong governmental visibility, active media presence);
- Moderate citizen engagement (limited participatory platforms);
- Low feedback integration (absence of systematic monitoring mechanisms).

A nationwide survey conducted by *Institut National de l'Environnement* (2022) found that 63% of Algerians consider media reports credible, but only 21% feel they can influence environmental decisions. This underscores a participation gap between institutional outreach and civic empowerment.

4.7 Proposed Framework: The Communicative Governance Cycle

To address these limitations, the study proposes the Communicative Governance Cycle (CGC)—a practical framework for embedding communication within every stage of environmental decision-making.

Table 7. The Communicative Governance Cycle (CGC)

Stage	Purpose	Key Activities	Tools and Mechanisms	Indicators of Success
Information	Ensure transparency and awareness	Public reports, press briefings, open data	Government portals, APS news, dashboards	Access frequency, information accuracy
Dialogue	Facilitate participatory discussion	Stakeholder forums, local consultations	Public hearings, radio debates	Number of participatory sessions, inclusivity index
Co-Creation	Joint development of solutions	NGO–state partnerships, media projects	Collaborative campaigns, educational programs	Number of joint initiatives, diversity of actors
Feedback	Evaluate impact and adjust policies	Citizen surveys, digital feedback tools	Online polls, SMS hotlines	Rate of policy adjustments, citizen satisfaction

This cyclical model transforms communication from a *reactive function* into a *strategic governance process*, fostering accountability and collective learning.

4.8 Quantitative and Qualitative Indicators

The following indicators can help assess Algeria's progress toward communicative sustainability:

Indicator Type	Measurement Dimension	Example
Quantitative	Media frequency index	Number of environmental articles or programs per month
Quantitative	Participation rate	Percentage of citizens engaged in local environmental initiatives
Qualitative	Message diversity	Inclusion of local languages, gender perspectives, and expert voices
Qualitative	Feedback responsiveness	Evidence of policy modification following public consultation

4.9 Challenges and Opportunities

Despite structural constraints, Algeria possesses considerable assets to advance communicative sustainability:

- **Expanding digital literacy**, especially among youth;
- **Institutional experience** in environmental awareness;
- **Rich cultural heritage** linking ecology with moral and religious values;
- **Active regional cooperation** through the African Union and UNEP initiatives.

Yet, persistent barriers remain:

- Bureaucratic inertia,
- Fragmented coordination between ministries and local governments,
- Insufficient research–policy integration, and
- Dependence on donor-funded NGO activities.

The opportunity lies in establishing a National Observatory for Environmental Communication, tasked with:

- Standardizing environmental reporting,
- Training journalists and local leaders,
- Monitoring communicative indicators aligned with the SDGs.

4.10 Conclusion

The analysis in this chapter demonstrates that Algeria's environmental governance is as communicative as it is administrative.

The effectiveness of sustainability policies depends on the quality of dialogue among institutions, citizens, and the media.

By institutionalizing communicative governance, Algeria can transition from fragmented awareness campaigns to a coherent participatory system, grounded in transparency, inclusion, and reflexivity. This would not only improve policy effectiveness but also strengthen civic trust and environmental democracy.

Chapter 5 – Rethinking Environmental Communication – Strategic Pathways for Algeria's Sustainable Future

5.1 Introduction

In the unfolding landscape of environmental governance, communication has moved from the margins of awareness campaigns to the very center of sustainable transformation.

For Algeria, this shift is both a necessity and an opportunity: a necessity because climate pressures—ranging from recurrent wildfires to prolonged droughts—demand coordinated societal responses; and an opportunity because the nation's evolving digital and cultural infrastructure enables unprecedented forms of dialogue and participation.

This chapter reimagines environmental communication as a strategic, systemic process underpinning Algeria's Vision 2030 for sustainable development.

It proposes that the next stage of progress will depend not on new technologies or isolated policies, but on the quality of communication linking state institutions, scientific expertise, and civil society.

5.2 The Global Transformation of Environmental Communication

Across the world, the philosophy of environmental communication has undergone a profound transformation.

What began in the 1970s as a branch of environmental education concerned primarily with raising awareness, has since evolved into a sophisticated interdisciplinary field centered on participation, governance, and co-creation of meaning.

Global experience demonstrates that sustainability cannot be achieved through top-down messaging. Instead, it requires the cultivation of communicative ecosystems—networks of mutual listening and shared learning that allow communities to make sense of ecological realities and act collectively.

In the European Union, this approach has materialized through the *Green Deal's Climate Pact*, which treats citizens as policy partners, not mere recipients.

In Latin America, participatory radio has enabled marginalized populations to shape local adaptation strategies. And in parts of Africa, digital storytelling has become a tool for collective ecological identity formation.

For Algeria, these transformations hold valuable lessons: effective environmental communication is not a matter of message volume but of message interaction—the ability to connect knowledge, trust, and social mobilization.

5.3 Algeria's Current Trajectory: Achievements and Gaps

Algeria has made visible progress in institutionalizing environmental awareness. Governmental agencies, notably the Ministry of Environment and Renewable Energies, have integrated communication into their national strategies.

Television, radio, and print media regularly cover environmental topics—from waste management to renewable energy—and educational programs in schools now include basic ecological literacy.

Yet, despite this visible presence, the communication system remains predominantly descriptive rather than dialogic.

Most public campaigns convey information about environmental hazards or state initiatives, but rarely open spaces for community deliberation or citizen feedback. This imbalance mirrors a broader global pattern in which communication is treated as an instrument of persuasion rather than a platform for participation.

Empirical analyses of Algerian media show that environmental reporting is often event-based—centered on commemorations such as "Environment Day" or emergency coverage of forest fires—while systemic, long-term issues like land degradation or water governance receive limited analytical depth.

The result is a form of "episodic awareness": citizens are informed, but not engaged; aware, but not empowered.

5.4 The Urgency of a Communicative Turn

To move beyond this stage, Algeria must undertake what scholars call a communicative turn in sustainability governance—a paradigm shift that treats communication not as a support activity but as a foundational process of environmental decision-making.

This transformation begins with acknowledging that sustainable policies succeed only when they are socially understood and collectively owned.

The Ministry of Environment's national surveys (2023) reveal that although over two-thirds of Algerians express concern about environmental degradation, fewer than one-third participate in local initiatives.

This gap between awareness and engagement points to a failure not of will, but of communication: messages circulate widely but fail to transform into shared responsibility.

Bridging this divide requires moving from the linear transmission model—where institutions inform citizens—to a reflexive communication model, where information flows in both directions, allowing citizens to question, adapt, and co-produce environmental solutions.

5.5 Strategic Directions for Communicative Sustainability

A coherent strategy for the next decade should rest on five interlinked priorities, each responding to a structural weakness in Algeria's current system while aligning with global standards.

1. Institutional Integration

Communication must be embedded within the architecture of environmental policy itself.

This means creating permanent coordination mechanisms among ministries, public media, and civil society to harmonize environmental narratives.

A **National Observatory for Environmental Communication** could serve as a monitoring hub, collecting data, evaluating public campaigns, and ensuring that communication indicators form part of Algeria's reporting on the Sustainable Development Goals (SDGs).

2. Professionalization of Environmental Journalism

Effective governance requires specialized mediators capable of translating complex environmental data into accessible narratives.

Algeria needs university programs and continuing education tracks that prepare journalists, communicators, and local officials for sustainability-oriented reporting.

Rather than viewing the environment as a "soft" topic, newsrooms must recognize its political and economic relevance, linking it to agriculture, health, energy, and tourism.

3. Digital Transformation and Open Data

The future of environmental communication lies in transparency and digital accessibility.

Establishing open data portals would allow citizens, researchers, and journalists to access real-time information on water quality, air pollution, and forest coverage. Mobile applications could enable community reporting of environmental incidents, while AI-based sentiment analysis tools might help policymakers gauge public perception and misinformation trends.

In short, data must become dialogue—an invitation to participation rather than a static archive.

4. Community and Cultural Anchoring

Beyond institutions and technologies, sustainability is a cultural process.

In Algeria's rural and Amazigh regions, environmental stewardship is deeply embedded in traditions of cooperation and shared responsibility (*touiza*).

Integrating these practices into environmental communication programs can transform abstract global narratives into locally resonant values.

Collaborations with religious institutions can also reinforce the moral dimension of ecological care, drawing on Islamic principles of *khilāfa* (stewardship) and *amāna* (trust).

5. Regional and International Partnerships

Finally, Algeria's communicative sustainability should be positioned within broader continental and global networks.

Participation in initiatives such as the African Union's Agenda 2063 and UNEP's MENA Communication Program would facilitate knowledge exchange, research collaboration, and the creation of joint media content addressing transboundary environmental issues.

These five pathways, if pursued coherently, can transform environmental communication from a fragmented practice into a systemic framework for participatory governance.

5.6 Projected Impact: From Awareness to Empowerment

The adoption of a communicative sustainability approach would yield both tangible and intangible benefits.

On the institutional level, it would enhance policy coherence, reduce duplication, and foster trust between government agencies and citizens. Socially, it would promote collective ownership of sustainability agendas, as communities begin to see environmental protection not as an external obligation but as an intrinsic civic value. Digitally, it would position Algeria among emerging nations that leverage open communication ecosystems for green innovation and accountability.

The ultimate goal is to create what might be called a culture of sustainable communication—a culture in which transparency, participation, and dialogue are habitual practices rather than exceptional efforts.

5.7 Challenges and Ethical Imperatives

No transformation occurs without resistance. Entrenched bureaucratic structures, limited funding, and uneven digital access pose significant challenges.

Equally important is the ethical dimension: communication must not become another tool of propaganda, but rather a medium of empowerment grounded in truth, inclusion, and responsibility.

This ethical commitment—rooted in transparency and cultural authenticity—will determine whether Algeria's communication system evolves into a participatory model or remains a rhetorical instrument.

5.8 Conclusion

Rethinking environmental communication in Algeria thus means rethinking governance itself.

The environmental crisis is not only a matter of policy failure but also of communicative failure: a disconnection between knowledge, responsibility, and collective action.

By embracing communicative sustainability as a guiding principle.

Algeria can transition from informing to involving, from awareness to empowerment, and from isolated initiatives to a national dialogue of sustainability.

The success of this transition will depend not merely on budgets or technologies, but on a new communicative ethos—one that listens, learns, and acts in concert with the people it serves.

General Conclusion

The reflections developed throughout this book have shown that environmental communication is not a secondary instrument of environmental management but a fundamental condition of sustainability itself. Through the analysis of Algeria's environmental policies, media practices, and community initiatives, the book has argued that *communication is the invisible infrastructure of sustainable development—* the system through which ideas circulate, trust is built, and collective action becomes possible.

From its theoretical premises to its practical applications, the book has pursued a single guiding question:

How can communication transform sustainability from a political slogan into a living social process within the Algerian context?

The answer, as demonstrated across five chapters, lies in the integration of communicative principles—transparency, participation, reflexivity, and cultural relevance—into environmental governance.

1. From Awareness to Interaction

The first two chapters established the epistemological foundation of environmental communication, tracing its emergence as a distinct field at the intersection of ecology, sociology, and media studies.

The discussion emphasized that information alone does not generate behavioral change; only through dialogue and feedback can knowledge become empowerment.

This insight dismantles the traditional notion of communication as a one-way channel and redefines it as a *reciprocal social process* that sustains cooperation between citizens and institutions.

2. Communication as the Heart of Sustainability

The theoretical exploration in Chapter 3 demonstrated that sustainability is, at its core, **a** communicative act—a shared negotiation of values, meanings, and responsibilities. The concept of *communicative sustainability* proposed in this work situates dialogue and participation as the ethical and practical pillars of sustainable governance. It recognizes that ecological transitions succeed only when they are *discursively legitimized* within local cultures and *symbolically anchored* in collective imagination.

By connecting global paradigms with Algerian realities, the book articulated a hybrid framework capable of bridging the institutional and the communal, the technical and the cultural, the local and the global.

3. Algeria's Path Toward Communicative Governance

The empirical and policy-oriented analysis in Chapter 4 highlighted that Algeria has achieved significant progress in environmental awareness and institutional communication.

National strategies, public media, and local NGOs have contributed to a growing public consciousness of environmental risks and responsibilities.

Yet, the findings also revealed the persistence of a participation gap: citizens are aware of sustainability issues but remain peripheral to decision-making processes.

This diagnosis underscores the urgency of transforming environmental governance into a communicative governance, where information flows horizontally and citizens are treated as co-creators of environmental solutions rather than passive recipients of policy messages.

4. A Vision for 2030: Toward a Culture of Sustainable Communication

The final chapter articulated a forward-looking vision for Algeria's environmental future—one that places communication at the center of its sustainable development strategy.

It proposed a set of interdependent reforms: institutional integration, professionalization of environmental journalism, digital transparency, cultural localization, and regional cooperation.

Together, these pathways form a roadmap for 2030, designed to move Algeria from *environmental awareness* to *environmental empowerment*.

They envision a society where sustainability is not only a governmental objective but a shared cultural ethos, cultivated through education, media, and civic participation.

5. Theoretical and Practical Contributions

From an academic perspective, this book contributes to the international literature by formulating the notion of communicative sustainability as a theoretical lens applicable to developing contexts.

It demonstrates how the communicative turn—long discussed in Western scholarship—can be localized and adapted to North African societies with distinct cultural, linguistic, and institutional configurations.

Practically, the research offers a policy-oriented framework for integrating communication into Algeria's environmental governance architecture.

It provides concrete mechanisms—observatories, training programs, open data platforms, and community storytelling initiatives—that can translate discourse into institutional reform.

6. A New Paradigm for Algeria and Beyond

At its deepest level, the book advances a broader philosophical claim: that sustainability is not merely about conserving nature, but about reconstructing the communicative relationship between humans and their environment.

It calls for a new ecological rationality—one that listens as much as it speaks, that understands sustainability as a dialogue between generations, and that measures progress not only in hectares reforested or tons recycled, but in *trust rebuilt* and *voices included.*

For Algeria, this means cultivating a culture of sustainable communication—a culture in which information is transparent, participation is inclusive, and governance is reflexive.
For the broader Arab and African contexts, it offers a transferable model capable of harmonizing traditional knowledge, modern technology, and democratic dialogue.

7. Final Reflection

In the end, the greatest challenge of sustainability is not technical—it is communicative.

It is the challenge of creating a shared language through which people, institutions, and ecosystems can coexist and evolve together.

If Algeria succeeds in building such a communicative foundation, it will not only safeguard its natural heritage but also pioneer a new form of ecological citizenship, grounded in dialogue, ethics, and solidarity.

References

Al'Awadhi, 'A. (1992). Environmental Awareness and Social Communication. Cairo: League of Arab States.

Al-'Awadhi, 'A. (1996). Toward a Successful Environmental Communication Policy. Cairo: League of Arab States & UNDP.

Al-Dulaimi, 'A. R. (2012). Media of Development: Concepts and Approaches. Amman

Al-Dulaimi, A. R. (2012). *Environmental Values in Islamic Thought: The Concept of Stewardship and Human Responsibility.* Amman: Dar Al-Nafaes.

Algerie Presse Service (APS). (2023). *Rapports annuels sur la couverture médiatique environnementale 2020–2023.* Alger: APS Archive.

Anderson, A. (2014). *Media, Environment and the Network Society.* Palgrave Macmillan. https://doi.org/10.1057/9781137314086

Beck, U. (1992). *Risk Society: Towards a New Modernity.* London: Sage Publications.

Beltrán, L. R. (1976). Alien Premises, Objects, and Methods in Latin American Communication Research. *Communication Research, 3*(2), 107–134. https://doi.org/10.1177/009365027600300201

Bessaoud, O. (2013). *Développement rural et politiques agricoles en Algérie: Les défis de la durabilité.* Paris: CIHEAM.

Bourdieu, P. (1991). *Language and Symbolic Power.* Cambridge: Polity Press.

Boyd, E., & Jasanoff, S. (2022). *Reimagining Climate Governance.* Cambridge University Press. https://doi.org/10.1017/9781108991370

Boykoff, M. T. (2019). *Creative (Climate) Communications: Productive Pathways for Science, Policy and Society.* Cambridge University Press. https://doi.org/10.1017/9781108297878

Carvalho, A. (2007). Ideological Cultures and Media Discourses on Scientific Knowledge: Re-reading News on Climate Change. *Public Understanding of Science, 16*(2), 223–243. https://doi.org/10.1177/0963662506066775

Cox, R., & Pezzullo, P. C. (2021). *Environmental Communication and the Public Sphere* (6th ed.). Thousand Oaks, CA: SAGE Publications.

Craig, R. T. (1999). Communication theory as a field. *Communication Theory, 9*(2), 119–161. https://doi.org/10.1111/j.1468-2885.1999.tb00355.x

Dryzek, J. S. (2020). *The Politics of the Anthropocene.* Oxford University Press. https://doi.org/10.1093/oso/9780198809616.001.0001

Escobar, A. (1995). *Encountering Development: The Making and Unmaking of the Third World.* Princeton University Press. https://doi.org/10.1515/9781400821464

European Union. (2020). *The European Green Deal.* Brussels: European Commission. https://eur-lex.europa.eu/legal-content/EN/TXT/?uri=CELEX%3A52019DC0640

FAO. (2022). *Rural Development Report 2022: Building Resilient Rural Communities.* Rome: Food and Agriculture Organization of the United Nations. https://www.fao.org/publications

Fairclough, N. (1995). *Critical Discourse Analysis: The Critical Study of Language.* London: Longman.

Foucault, M. (1972). *The Archaeology of Knowledge.* New York: Pantheon Books.

Freire, P. (1970). *Pedagogy of the Oppressed.* New York: Continuum.

Giddens, A. (2009). *The Politics of Climate Change.* Cambridge: Polity Press.

Habermas, J. (1984). *The Theory of Communicative Action, Vol. 1: Reason and the Rationalization of Society.* Boston: Beacon Press.

Hajer, M. A. (1995). *The Politics of Environmental Discourse: Ecological Modernization and the Policy Process.* Oxford: Clarendon Press.

Hall, S. (1997). *Representation: Cultural Representations and Signifying Practices.* London: Sage Publications.

Hanawi, 'I. (2004). Environmental Issues in 100 Questions and Answers. Beirut: Environment and Development Publications.

Hansen, A. (2010). *Environment, Media and Communication.* London: Routledge. https://doi.org/10.4324/9780203860014

Hansen, A., & Machin, D. (2019). *Media and Communication Research Methods* (2nd ed.). London: Palgrave Macmillan. https://doi.org/10.1007/978-1-137-51283-0

Hess, D. J. (2016). *Undone Science: Social Movements, Mobilized Publics, and Industrial Transitions.* Cambridge, MA: MIT Press. https://doi.org/10.7551/mitpress/9780262034358.001.0001

Institut National de l'Environnement (INE). (2022). *Enquête nationale sur la perception publique des politiques environnementales en Algérie.* Alger: Ministère de l'Environnement.

Kalin, A. (2013). Ecological Communication Theory: Revisiting Systems Thinking. *Environmental Communication, 7*(4), 555–573. https://doi.org/10.1080/17524032.2013.816327

Killingsworth, M. J., & Palmer, J. S. (1992). *Ecospeak: Rhetoric and Environmental Politics in America.* Carbondale: Southern Illinois University Press.

Lerner, D. (1958). *The Passing of Traditional Society: Modernizing the Middle East.* New York: The Free Press.

Lester, L., & Hutchins, B. (2013). *Environmental Conflict and the Media.* London: Palgrave Macmillan. https://doi.org/10.1057/9781137273358

Manyozo, L. (2012). *Media, Communication and Development: Three Approaches.* New Delhi: Sage Publications. https://doi.org/10.4135/9788132113329

McCombs, M., & Shaw, D. (1972). The agenda-setting function of mass media. *Public Opinion Quarterly, 36*(2), 176–187. https://doi.org/10.1086/267990

Meadowcroft, J. (2011). Engaging with the Politics of Sustainability: From Ecological Modernization to Reflexive Governance. *Environmental Politics, 20*(1), 1–21. https://doi.org/10.1080/09644016.2011.536865

Melha, S. (2000). *Les politiques de communication en Algérie: De la centralisation à la participation.* Alger: Office des Publications Universitaires.

Ministry of Environment and Renewable Energies. (2023). *Rapport national sur la stratégie de développement durable et la communication environnementale en Algérie.* Algiers: Government of Algeria. *(Physical publication; see UNEP 2023 for digital alternative.)*

Mol, A. P. J., & Spaargaren, G. (2000). Ecological Modernization Theory in Debate: A Review. *Environmental Politics, 9*(1), 17–49. https://doi.org/10.1080/09644010008414511

Nisbet, M. C. (2009). Communicating climate change: Why frames matter for public engagement. *Environment: Science and Policy for Sustainable Development, 51*(2), 12–23. https://doi.org/10.3200/ENVT.51.2.12-23

Pezzullo, P., & Cox, R. (2021). *Environmental Communication and the Public Sphere* (6th ed.). Thousand Oaks, CA: Sage Publications.

Rachwan, M. (2006). Environmental Communication in Algeria: Realities and Prospects for Sustainable Development. Algiers: Université d'Alger.

Rachwan, ʿA. H. (2006). *ʿIlm al-ijtimaʿ al-bīʾī (علم الاجتماع البيئي)* [Environmental Sociology]. Alexandria: Al-Maktab al-Jāmiʿī al-Ḥadīth.

Rogers, E. M. (1962). *Diffusion of Innovations.* New York: Free Press.

Said, E. W. (1978). *Orientalism.* New York: Pantheon Books.

Schäfer, M. S., & Painter, J. (2023). *Climate Communication in the Digital Era.* Oxford University Press. https://doi.org/10.1093/oso/9780198880684.001.0001

Servaes, J. (1999). *Communication for Development: One World, Multiple Cultures.* Cresskill, NJ: Hampton Press.

Servaes, J. (2018). *Sustainable Development and Green Communication: African and Global Perspectives.* Singapore: Palgrave Macmillan. https://doi.org/10.1007/978-981-10-8591-9

UNDP. (2019). *Guidelines on Communication for Sustainable Development (C4D).* New York: United Nations Development Programme. https://www.undp.org/arab-states

UNDP. (2022). *Evaluation of Environmental Communication Practices in North Africa: Case Studies from Algeria and Tunisia.* Cairo: UNDP Regional Office for Arab States. https://www.undp.org/arab-states

UNEP. (2023). *Environmental NGOs and Civil Society Participation in the MENA Region: Annual Report.* Nairobi: United Nations Environment Programme. https://www.unep.org/resources/publications

UNESCO. (2021). *Education for Sustainable Development: Learning for Our Planet, Learning for Sustainability.* Paris: UNESCO. https://unesdoc.unesco.org/ark:/48223/pf0000376707

World Bank. (2024). *Green Growth and Public Participation in North Africa: Governance Challenges and Communication Strategies.* Washington, DC: World Bank Publications. https://documents.worldbank.org

www.ingramcontent.com/pod-product-compliance
Lightning Source LLC
LaVergne TN
LVHW022310200926
848481LV00004B/1271

* 9 7 9 8 8 9 9 6 6 2 5 0 8 *